T0212126

History of Analytic Philosophy

Series Editor: **Michael Beaney**

Titles include:

Forthcoming:

Sébastien Gandon
RUSSELL'S UNKNOWN LOGICISM
A Study in the History and Philosophy of Mathematics

Anssi Korhonen
LOGIC AS UNIVERSAL SCIENCE
Russell's Early Logicism and Its Philosophical Context

Consuelo Preti
THE METAPHYSICAL BASIS OF ETHICS
The Early Philosophical Development of G.E. Moore

Sandra Lapointe *(translator)*
Franz Prihonsky
THE NEW ANTI-KANT

Erich Reck *(editor)*
THE HISTORIC TURN IN ANALYTIC PHILOSOPHY

Maria van der Schaar
G.F. STOUT: ON THE PSYCHOLOGICAL ORIGIN OF ANALYTIC PHILOSOPHY

Pierre Wagner *(editor)*
CARNAP'S IDEAL OF EXPLICATION AND NATURALISM

History of Analytic Philosophy

Series Standing Order ISBN 978–0–230–55409–2 (hardcover)

Series Standing Order ISBN 978–0–230–55410–8 (paperback)

outside North America only

You can receive future titles in this series as they are published by placing a standing order. Please contact your bookseller or, in case of difficulty, write to us at the address below with your name and address, the title of the series and the ISBN quoted above.

Customer Services Department, Macmillan Distribution Ltd, Houndmills, Basingstoke, Hampshire RG21 6XS, England

Also by Douglas Patterson

NEW ESSAYS ON TARSKI AND PHILOSOPHY *(editor)*

Alfred Tarski: Philosophy of Language and Logic

Douglas Patterson
Visiting Scholar, University of Pittsburgh, Pittsburgh, PA

Softcover reprint of the hardcover 1st edition 2012 978-0-230-22121-5

First published 2012 by
PALGRAVE MACMILLAN

Palgrave Macmillan in the UK is an imprint of Macmillan Publishers Limited,
registered in England, company number 785998, of Houndmills, Basingstoke,
Hampshire RG21 6XS.

Palgrave Macmillan in the US is a division of St Martin's Press LLC,
175 Fifth Avenue, New York, NY 10010.

Palgrave Macmillan is the global academic imprint of the above companies
and has companies and representatives throughout the world.

Palgrave® and Macmillan® are registered trademarks in the United States,
the United Kingdom, Europe and other countries

ISBN 978-1-349-30673-2   ISBN 978-0-230-36722-7 (eBook)
DOI 10.1057/9780230367227

Contents

Series Editor's Foreword

During the first half of the twentieth century analytic philosophy gradually established itself as the dominant tradition in the English-speaking world, and over the last few decades it has taken firm root in many other parts of the world. There has been increasing debate over just what 'analytic philosophy' means, as the movement has ramified into the complex tradition that we know today, but the influence of the concerns, ideas and methods of early analytic philosophy on contemporary thought is indisputable. All this has led to greater self-consciousness among analytic philosophers about the nature and origins of their tradition, and scholarly interest in its historical development and philosophical foundations has blossomed in recent years. The result is that history of analytic philosophy is now recognized as a major field of philosophy in its own right.

The main aim of the series in which the present book appears – the first series of its kind – is to create a venue for work on the history of analytic philosophy, consolidating the area as a major field of philosophy and promoting further research and debate. The 'history of analytic philosophy' is understood broadly, as covering the period from the last three decades of the nineteenth century to the start of the twenty-first century – beginning with the work of Frege, Russell, Moore and Wittgenstein, who are generally regarded as its main founders, and the influences upon them – and going right up to the most recent developments. In allowing the 'history' to extend to the present, the aim is to encourage engagement with contemporary debates in philosophy – for example, in showing how the concerns of early analytic philosophy relate to current concerns. In focusing on analytic philosophy, the aim is not to exclude comparisons with other – earlier or contemporary – traditions, or consideration of figures or themes that some might regard as marginal to the analytic tradition but which also throw light on analytic philosophy. Indeed, a further aim of the series is to deepen our understanding of the broader context in which analytic philosophy developed, by looking, for example, at the roots of analytic philosophy in neo-Kantianism or British idealism, or the connections between analytic philosophy and phenomenology, or discussing the work of philosophers who were important in the development of analytic philosophy but who are now often forgotten.

In this book Douglas Patterson provides the first full-length account of Alfred Tarski's philosophy. Tarski was born in Warsaw in 1901 and gained his doctorate in logic at the University of Warsaw in 1924, supervised by Stanisław Leśniewski. In the 1920s and 1930s he published extensively on logic and set theory, and as a representative of the so-called Lvov–Warsaw School, maintained close links with Gödel, Carnap and other members of the Vienna Circle. When Nazi Germany invaded Poland on 1 September 1939, Tarski was at a conference in the United States and was unable to return home. He stayed there throughout the war years, teaching at Harvard, New York and Princeton before eventually being given a permanent post at the University of California at Berkeley in 1945, where he remained until his death in 1983. Tarski was thus one of the many logicians and philosophers from Central Europe who moved to the United States as a result of the rise of Nazism in Germany, and whose story is part of the broader story of the development of analytic philosophy in North America, as the ideas of the Polish logicians and the logical positivists took root in new soil.

Tarski's two most famous papers are 'The Concept of Truth in Formalized Languages', first published in Polish in 1933 (and in German in 1935) and 'On the Concept of Logical Consequence', published in both Polish and German in 1936. In the first paper Tarski offers a definition of truth for formal languages by introducing the notion of satisfaction and appealing to the recursive structure of formal languages. Tarski first formulates his 'T-schema' here (famously exemplified in the statement that 'Snow is white is true' if and only if snow is white), and also offers a solution to the Liar paradox by insisting that truth for a language can only be defined in a metalanguage. In the second paper Tarski offers a corresponding semantic definition of logical consequence by utilizing the idea of truth-under-an-interpretation introduced in his earlier paper. These two papers have been seen as establishing the foundations of truth-conditional semantics, with Tarski also regarded as a key figure in the development of model theory.

As Patterson argues in this book, however, Tarski's actual views are both more complex and more intriguing than they have standardly been taken to be. Patterson begins by distinguishing what he calls the 'expressive' conception of meaning from the 'representational' conception of meaning. On the expressive conception, language expresses thoughts and the notions of assertion and justification have primacy. On the representational conception, language represents things in the world and the notions of reference and truth have primacy. The former finds its natural home in the proof-theoretic conception of logic, while the latter is reflected in the model-theoretic conception of logic. Patterson agrees that Tarski's two papers made major contributions to representational semantics and model theory, but by carefully examining the development of Tarski's work from the late 1920s to the mid-1930s, he shows that that work was originally motivated by the expressive conception of meaning. Tarski himself referred to his earlier view as 'intuitionistic formalism', a view that he inherited from Leśniewski. It was only once he had convinced himself that logical consequence could be defined semantically that he abandoned his earlier view. Patterson remarks in his introduction that "Tarski may well be the most enigmatic figure in the history of analytic philosophy". Patterson's detailed account of the crucial period in Tarski's intellectual development not only sheds a great deal of light on Tarski's evolving ideas on the fundamental notions of logic and semantics but also inaugurates a new era in our understanding of Tarski's work and its contribution to analytic philosophy.

Michael Beaney
September 2011

Introduction

Expressive and representational semantics

This book tells the story of the birth of truth-conditional semantics from an earlier conception of meaning. If we think of language as standing between mind and world, there are two simple ways in which to think of it as meaningful: in terms of its relation to the mind, or in terms of its relation to the world. One may thus conceive of meaning in terms of the expression of thought, or in terms of the representation of things. Call the two conceptions *expressive* and *representational* semantics. On the expressive conception the function of language is to express thoughts, which may themselves be representational. On the representational conception language is conceived of as representational in its own right, and the expression of thought is a derivative function. Central to the expressive conception of language are the notions of assertion and justification since our basic notions are a subject's saying something and their reasons for doing so; central to the representational conception are the notions of reference and truth being about things and accurately representing how they are.

Both views address the topic of inference. The expressive conception offers a natural, intuitive connection to the basic idea of an argument as something that gives one a reason to believe one thing given that one believes others: in a valid argument, justification for believing the premises is transmitted to the conclusion and becomes justification for believing it. The representational conception, since its focus is on accuracy in what statements are about, gives rise to the idea that an argument is valid just in case if its premises are true, so is its conclusion. Continuing in the same broad brush-strokes, the expressive conception sits naturally with the conception of inference as derivation of one claim from others

1

in accord with intuitively valid rules—a conception that becomes, in more refined studies of logic, proof theory. The representational conception naturally leads to the idea of an inference as valid just in case all models of the premises are also models of the conclusion, an idea at the foundation of model-theoretic studies of logic. The interaction of the two conceptions in their refined forms then gives us two of the central results of the 20th century, Gödel's completeness and incompleteness theorems.

Neither conception of meaning is sufficient of itself; likewise, neither the proof-theoretic nor the model-theoretic conception of logic of itself captures the notion of one thing's following from another. Indeed, when it comes to logic, both derivation and semantic consequence are arguably incomplete or even inaccurate representations of what it is for one claim to follow from others. Derivational conceptions of consequence seem limited to the finite, and there isn't anything particularly explicit in the "therefore" of ordinary deductive argumentation that concerns rules of proof, while semantic conceptions of consequence seem to lose the epistemic cast of this same "therefore", since one thing might be the case if others were even if their being so gave one, even in principle, no reason at all to think that it was so. The relations among the intuitive notions of deductive validity, of derivation, and of semantic consequence will be a matter for further investigation as we go along. However, whatever the consequences of that investigation might be, the refinement of these two traditions into proof theory and model theory was a major achievement.

Alfred Tarski is remembered by students of language for contributions that made rigorous treatments of the central concepts of the second of these two traditions possible. In "The Concept of Truth in Formalized Languages" (CTFL) and related papers in the early 1930s Tarski developed the tools and techniques that are still at the center of logic, the philosophy of language and, to a lesser extent, linguistic semantics. Indeed, Tarski's achievements in this regard were so influential that today it is forgotten that representational semantics as he developed it was devised as a contribution to a certain project motivated by the expressive conception. Standard reports on Tarski's work simply treat it as an obvious early contribution to representational semantics. Lurking in the text of the classic papers, however, is a different project, one to which representational semantics was at first intended as a small contribution. Little is known about this project today other than that at one place [Tarski, 1983a, 62] Tarski refers to the view he was working with as "Intuitionistic Formalism".

In this work we will examine the development of truth-conditional, representational semantics by Tarski from the late 1920s until the mid-1930s. Three basic foci of attention will allow us to fill in the surrounding picture from which representational semantics is usually anachronistically abstracted. First, we will examine what Intuitionistic Formalism was, where it came from, and how Tarski conceived of himself as contributing to it. Second, we will look at the development of semantics from its beginnings within Intuitionistic Formalism to its replacement of it in 1935. Finally, we will trace Tarski's treatment of logical consequence through its development in his thinking, for it was when Tarski realized that semantics as he had developed it could replace the conception of inference required by Intuitionistic Formalism that the latter view itself lost its hold on him.

Our story, summarized, goes like this. Tarski originally became interested in the question of how our thoughts and ideas, under a certain conception of what those are, could adequately be expressed in an axiomatic theory or "deductive science", a topic that was central to his advisor Stanisław Leśniewski's work. As Intuitionistic Formalism was conceived of by Leśniewski as a conception of the function and significance of an axiomatic theory, Tarski set himself the task of exploring how the basic concepts used in thought *about* axiomatic theories—consequence, truth, reference and related notions such as completeness and categoricity (in several senses)—could themselves be captured, to Intuitionistic Formalist standards, within an axiomatic theory. In particular due to his interest in early work in what we would now think of as model theory, Tarski set out to develop a way of capturing the semantic notions of truth, satisfaction and reference within such a theory to Intuitionistic Formalist standards. The result of this project was the now-familiar method of defining truth by recursion on satisfaction.

Logical consequence, on the other hand, was much slower to develop in Tarski's hands. Aside from a skeptical footnote following Gödel [Tarski, 1983a, 252], as late as 1934 Tarski treated logical consequence derivationally, in terms of a recognized set of apparently valid rules for asserting sentences given that others had been asserted. Only when he realized that the notion of a model could be defined in terms of semantics, and consequence in turn defined in terms of that, did Tarski see that semantics could actually stand on its own as a treatment of language, and at that point he moved on from the project that had originally motivated him: after 1936 Intuitionistic Formalist concerns disappear from his work and, in particular, 1944's "The Semantic Conception of Truth", though it appears to summarize the work of the 1930s, leaves out the themes characteristic of Intuitionistic Formalism.

The received view

Tarski may well be the most enigmatic figure in the history of analytic philosophy. Although his importance is widely accepted, his work is rarely actually read and interpretations and assessments of his views differ wildly. It was this way from the beginning. Popper first learned of Tarski's innovations when he asked Tarski to explain his work:

> and he did so in a lecture of perhaps twenty minutes on a bench (an unforgotten bench) in the *Volksgarten* in Vienna. He also allowed me to see the sequence of proof sheets of the German translation of his great paper on the concept of truth, which were then just being sent to him ... No words can describe how much I learned from all this, and no words can express my gratitude for it. Although Tarski was only a little older than I, and although we were, in those days, on terms of considerable intimacy, I looked upon him as the one man whom I could truly regard as my teacher in philosophy. I have never learned so much from anybody else [Popper, 1974, 399].

Carnap recalls his own initiation:

> When Tarski told me for the first time that he had constructed a definition of truth, I assmed that he had in mind a syntactical definition of logical truth or provability. I was surprised when he said that he meant truth in the customary sense, including contingent factual truth. Since I was only thinking in terms of a syntactical metalanguage, I wondered how it was possible to state the truth-condition for a simple sentence like "this table is black". Tarski replied: "This is simple: the sentence 'this table is black' is true if and only if this table is black" ... When I met Tarski again in Vienna I urged him to deliver a paper on semantics and on his definition of truth at the International Congress for Scientific Philosophy to be held in Paris in September. I told him that all those interested in scientific philosophy and the analysis of language would welcome this new instrument with enthusiasm, and would be eager to apply it in their own philosophical work. [Carnap, 1963, 60–1].

However, Tarski was skeptical, and rightly so. On the one hand, that "this table is black" is true if and only if this table is black can easily seem something less than a penetrating insight; on the other hand, as

Carnap's reference to syntax indicates, there was at the time a great deal of skepticism or outright hostility toward the treatment of language in terms of word–world relations. The latter reaction dominated the the Congress, where "there was vehement opposition even on the side of our philosophical friends" [Carnap, 1963, 61]. Posterity hasn't been more accommodating, though the sources of resistance have shifted. Consider these familiar assessments of Tarski's views:

> The concern of philosophy is precisely to discover what the intuitive notion of truth is. As a philosophical account of truth, Tarski's theory fails as badly as it is possible for an account to fail [Putnam, 1994, 333].

> My claim is that Tarski's analysis is wrong, that his account of logical truth and logical consequence does not capture, or even come close to capturing, any pretheoretic conception of its logical properties [Etchemendy, 1990, 6].

What provoked these disparate reactions? The standard conception of Tarski's views is relatively easy to summarize. In "The Concept of Truth in Formalized Languages" [Tarski, 1983a] (CTFL) Tarski sets himself the task of "defining truth" and maintains that a good definition of truth would be something that, added to the formal syntax of a language, results in implication of a sentence of the form "*s* is true if and only if p" (as we now say, a "T-sentence") for each sentence of the language, where what is substituted for "p" is or translates *s*. If this is the standard, Tarski notes, then if a language had a finite number of sentences one could simply list the T-sentences and be done with it [Tarski, 1983a, 188]. The technical achievement comes in Tarski's recognition that if a language has an infinite number of sentences and one demands a finite number of axioms in one's syntactic-cum-semantic theory, some use of "the recursive method" [Tarski, 1983a, 189] is required. Furthermore, since complex sentences aren't necessarily built out of sentences (e.g. $\exists x F x$ is built by concatenating the quantifier with the open "*Fx*") the basic recursive definition has to concern something other than closed sentences, but then produce, for closed sentences, a definition with the T-sentences as consequences.

The details here are familiar and so I refer the reader to standard presentations such as [Soames, 1999]. The basics are encountered in any advanced undergraduate logic class. Consider a language $\mathcal{L}$, with names, predicates and functors and variables of the first and perhaps higher orders and quantifiers to bind them, plus some stock of sentential connectives. Syntactically, one sets out rules for forming a sentence (open or closed) by stringing together names, functors, predicates, variables, quantifiers and connectives; some notational bookkeeping is done with devices like parentheses, the dots of the Peano school, or order alone, as in Polish notation, to determine things like the scope of a quantifier or the main connective of a sentence. The rules sort strings into formulas and non-formulas. Since they are recursive (e.g. one can conjoin a conjunction with something else or apply a functor to something that contains functors) the result is an infinite set of formulae, some open and some closed, where an open formula contains at least one variable bound by no quantifier.

Semantically, by Tarski's techniques, one assigns denotations to names and sets of ordered n-tuples to predicates and to functors. The former n-tuples represent objects as standing in the relation expressed by the predicate and the latter have a distinguished element (the last, say) that expresses the reference of the functor when it is applied to the first $n - 1$ members of the n-tuple as arguments. Truth-functions are assigned to the sentential connectives in the obvious way. Variables are enumerated and objects in the domain of discourse are taken to form sequences—enumerations of objects to correspond to enumerations of variables. A sentence s with a relation symbol R and no bound variables is satisfied by a sequence just in case the n-tuple of objects with indices that match those of the free variables in s is one of the n-tuples assigned to R. If the sentence involves a universal quantifier binding variable x_i it is satisfied by a sequence if and only if every sentence that differs from that sequence at most the x_i^{th} place satisfies the open sentence from which it was formed by adding the quantifier [Tarski, 1983a, 193]. Closed sentences being those with no free variables, they are satisfied by every sequence if they are satisfied by any, and so a true sentence is defined as one satisfied by every sequence [Tarski, 1983a, 194–5]. Given that reference and satisfaction have themselves been introduced by translational enumeration (e.g. "'Frankreich' refers in German to France") the result is an account that implies the T-sentences as desired.

As the standard conception notes, Tarski was motivated by the antinomy of the liar and related semantic paradoxes and took them to be the primary obstacle to a rigorous treatment of the concept of truth. These lead him to say that no language can contain a predicate applying to exactly the set of (Gödel codes of) its own truths and that therefore the definition of truth for a language can only be given in a "richer" metalanguage. The result, on the standard reading, is the "Tarski hierarchy" in which the semantics for a language must be given in metalanguage that cannot be translated into it, the semantics of which in turn must be stated in a yet higher "meta-metalanguage", and so on. Since natural languages appear to contain self-applying semantic vocabulary and this apparently violates the restrictions imposed by the hierarchy, Tarski condemns them as "inconsistent" and rejects them as unfit for rigorous study.

Continuing with the standard narrative, in 1935–6 Tarski turned his attention to the relation of logical consequence. Here he defines a model of a set of sentences as a sequence that satisfies all of them and then defines an argument (Γ, s) as valid just in case every model of Γ is a model of s. There are some oddities about Tarski's presentation, but it is widely held to be more or less the first statement of the usual contemporary conception of model-theoretic consequence.

As we have noted, reactions have been extreme. Neurath thought that Tarski was rehabilitating metaphysics and rejected the idea that the truth of a sentence could be understood in terms of worldly conditions rather than its confirmation within a system of scientific statements [Mancosu, 2008]. Later thinkers have questioned whether Tarski really provided a "reduction" of semantic notions to physicalistically acceptable terms [Field, 1972]. Interpreters have often objected to the apparent restriction of Tarski's "definition" of truth to a single language. Others have pilloried Tarski's remarks on natural language. Tarski's response to the paradoxes, more generally, is the main foil for contemporary work on the paradoxes, work that attempts to show that, contrary to his pronouncements, languages that contain their own truth-predicates can be constructed. As for Tarski's account of consequence, though it has found defenders, many agree with Etchemendy's assessment quoted above.

Themes

So much for the received view. For those keeping score, I will mostly disagree with the usual objections to Tarski's views on truth and paradox and mostly agree with the objections to Tarski's account of consequence. The details will emerge as we progress. But keep in mind that my goal here is to explain where Tarski's views came from and how they inform what he wrote at the time rather than to defend everything that he said.

The overall story here is a developmental one. Tarski begins in the late 1920s working within a conception of language and logic inherited from his mentors Stanislaw Leśniewski and Tadeusz Kotarbiński. He sets himself the task of working out "Intuitionistic Formalist" accounts of basic metatheoretical notions—in the first case, the consequence relation conceived of as determined by a set of primitively accepted rules of inference [Tarski, 1983c, 63]. Sometime in 1929 he turns his attention to semantic notions and develops the basic techniques described above, at first imperfectly in [Tarski, 1983f] and then in full form in [Tarski, 1983a]. During this period he doesn't think to apply the techniques to the consequence relation, nor does he notice other obvious applications, e.g. to Padoa's method for proving the independence of a set of axioms. These omissions are important because they show the extent to which Tarski is still thinking squarely in terms of Intuitionistic Formalism. Sometime in 1934–5 Tarski realized that he could replace the appeal to primitively valid inference rules with a notion of consequence defined in terms of his semantics. The result is the account of [Tarski, 2002] and with that the project came to an end.

As a bit of a preview, here are some themes that we will be exploring as we move along.

1. *Intuitionistic Formalism*: What was it and how did it inform Tarski's work from 1926–1936?
2. *Definition*: What was Tarski's conception of definition? How do his definitions relate to what philosophers offer when they "analyze" concepts?
3. *Language*: What were Tarski's views on Language? What was the implicit account of meaning and communication with which Tarski worked?
4. *Semantics*: What motivated Tarski's project of defining semantic terms in mathematical theories? How did he understand what he was doing at the time?
5. *Compositionality*: Intuitionistic Formalism not only lacked any articulate account of how the meanings of a complex expression were determined by those of its parts—something required of any reasonable treatment of language—but in Tarski's hands it involved an account of meaning that was incompatible with compositionality. We will see how Tarski's semantics begins to supply this. One of the aspects of this shift is the move from a "holist" account of meaning on which theories determined the meaning of their sentences and expressions, to an "atomist" one which the determination runs from lexical expressions to sentences.
6. *Consequence*: Intuitionistic Formalism involved a conception of logical consequence on which it was a matter of primitively valid rules of inference. Tarski's successor conception detaches consequence from this requirement. The result is that Tarski moves from a conception of consequence that places its epistemic aspects at the forefront to

one that stresses the independence of logic from the semantics of non-logical terms. Call the contrast one between "epistemic" and "generality" conceptions of consequence. Whether this is laudable is something we will consider at length.

7. *Logical Monism and Pluralism*: Tarski began his work accepting that Simple Type Theory (STT), conceived of in terms of the "theory of semantical categories" he attributed to Leśniewski and others, was logic. Under the influence of Carnap in particular, but also paying attention to other developments, he moved to a more pluralist or, as Carnap would put it, "tolerant" attitude.

8. *The Status of Mathematics*: Tarski inherited a liberal Polish attitude about the status of mathematics. Though he nowhere takes the logicist program as a motivation, a good deal of his views assume that mathematics, to the extent that it can be developed in STT, is logic. Later, especially beginning with [Tarski, 2002] a more open-minded attitude about logic and mathematics came to the fore.

Those, then, are some of the themes that we will explore. That said, I owe the reader some words about what will not be covered in this book. Our topic is Tarski's evolving set of views about logic and language, and in particular "formal axiomatics" in the period 1926–1936. Tarski's mathematical work from the period will not be covered.[1] Nor will I have space to discuss Tarski's views after 1936, with the exception of "The Semantic Conception of Truth", published in 1944 but largely a response to the debate that arose following his presentation of "The Establishment of Scientific Semantics" (ESS) and "On the Concept of Following Logically" (CFL) at the Paris Unity of Science Congress in 1935.[2]

As for other sources, in order to stay within the publisher's length limit I have set myself the basic rule of sticking to figures and works Tarski himself mentions, with only a few exceptions when these seemed especially pertinent. Missing, in particular, are the discussions of Russell and Frege that one might expect in a work on the history of analytic philosophy. Tarski was in a position to have received at least some Frege from Leśniewski, but Frege played no direct role in his thought. Russell seems relevant at the many places where Tarski mentions *Principia Mathematica*, but since Tarski mentions it only to set it aside in favor of STT, I found little of use in bringing Russell into our discussion. Wittgenstein will figure into our story only indirectly, as filtered through Carnap and the Vienna Circle. One comparison seemed too interesting to me to leave out, and so it occurs in foonotes: that of Tarski's work with contemporaneous writings of his unfortunately neglected colleague Kazimerz Ajdukiewicz.

There are also some topics that students of Tarski in particular might expect to see discussed more than they will be in this book. First, I give the distinction between language and metalanguage less attention than it often receives in the literature since it is perfectly clear and and there aren't any significant issues concerning it; interested readers can consult [Woleński, 1989] and [Betti, 2004, 280ff] on Leśniewski's development of the distinction. Second, I spend only a few words on the nominalism of Leśniewski and Kotarbiński, despite the fact that Tarski in private discussions expressed sympathy for it throughout his life, as is well known; here, too, interested readers can begin with [Woleński, 1989]. Others have written on the topic to a considerable extent but in my research I came to the conclusion that it is of little importance for our main themes.

This book is something of an exercise in philosophical detective work. Tarski is known for having been reticent in his expression of philosophical commitments. Even in the overtly philosophically motivated works his views are difficult to discern. The method I have followed is an obvious one: figure out what the views of his teachers were and then find sufficient signs of those views in what Tarski did say that we can feel comfortable filling in the gaps with those views. The proof is in the interpretations offered of familiar passages. I try to make a good case for my reading in chapters 4, 5 and 7.

A final word on textual interpretation is in order. Meticulous as he may have been in logic and mathematics, Tarski makes no effort to impose order on his use of philosophically loaded terminology. The eager interpreter hoping to make something of Tarski's varying uses of, in German, *Sinn*, *Bedeutung* and *Inhalt*, for instance, is quickly disappointed. Though the work of the 1930s purports to be concerned with "concepts" by 1944, in note 4 to SCT [Tarski, 1986e, 697], Tarski says that he has a tendency to avoid the word in "exact discussion", since on its prevalent usage it sometimes refers to extensions, sometimes to meanings, and sometimes to words themselves. He must have had his earlier self in mind in this passage, because all of these usages can be found in the work of the 1930s. Despite the fact that his views sometimes demand distinctions within the range of coverage of these and other philosophically fraught terms, in various passages Tarski uses one or the other of them indifferently; e.g. in one place [Tarski, 1983k, 311] [Tarski, 1986b, 647] Tarski talks about an axiom system determining the "senses" (*Sinn*, singular: the English translation substitutes the plural) of the "concepts" (*Begriffe*) it "contains"; several lines later "concepts" is traded in for "terms" (*Zeichen*). These difficulties are then compounded by various issues with the translations—for example the one just noted. Often this doesn't matter, but in some cases it does, as in the attempt to determine whether Tarski held that there was one concept of truth that applied to many languages, or rather for each language its own concept of truth. While trying to be as faithful as possible to the texts I sometimes allow my confidence in the overall reading to trump considerations of wording. This is a familiar dilemma facing the historian of philosophy. I hope that I have handled it well.[3]

1
Intuitionistic Formalism

1.1 What was Intuitionistic Formalism?

1.1.1 A puzzle about concepts and definitions

Tarski begins "The Concept of Truth in Formalized Languages" thus:

> The present article is almost wholly devoted to a single problem—
> *the definition of truth*. Its task is to construct—with reference to a given
> language—*a materially adequate and formally correct definition of the term*
> *'true sentence'* [Tarski, 1983a, 152].

Despite this desire to construct a definition, Tarski also insists that he
will not be "analyzing" the meaning of the word "true":

> A thorough analysis of the meaning current in everyday life of the
> term 'true' is not intended here. Every reader possesses in greater or
> less degree an intuitive knowledge of the concept of truth and he can
> find detailed discussions on it in works on the theory of knowledge
> [Tarski, 1983a, 153].

This raises a question. If the definition isn't intended to provide "a
thorough analysis of the meaning of the term 'true'", what is it for?
Furthermore, notice that what this passage says is that Tarski will *not*
be analyzing the meaning of "true" or the concept of truth; his work
assumes that this analysis is already available and can be found "in
works on the theory of knowledge"—in particular, we learn in a footnote, Kotarbiński's *Elementy* [Kotarbiński, 1966]. So whatever Tarski is
doing in "constructing a materially adequate and formally correct definition of truth", he is not engaged in "analysis" of the sort undertaken
by philosophers.

That Tarski has no intention of doing what philosophers do when they "define" or "analyze" concepts is borne out by the definitions he actually gives. The net effect of Definitions 22 and 23 of the work is certainly not to increase our confidence in some antecedent grasp we might have taken ourselves to have on the concept of truth:

> Definition 22. *The sequence f* satisfies *the sentential function x if and only if f is an infinite sequence of classes and x is a sentential function and if f and x are such that either* (α) *there exist natural numbers k and l such that* $x = \iota_{k,l}$ *and* $f_k \subseteq f_l$; (β) *there is a sentential function y such that* $x = \bar{y}$ *and f does not satisfy the function y;* (γ) *there are sentential functions y and z such that* $x = y + z$ *and f either satisfies y or satisfies z; or finally* (δ) *there is a natural number k and a sentential function y such that* $x = \bigcap_k y$ *and every infinite sequence of classes which differs from f in at most the k-th place satisfies the function y* [Tarski, 1983a, 193].

> Definition 23. *x is a* true sentence—*in symbols* $x \in Tr$—*if and only if* $x \in S$ *and every infinite sequence of classes satisfies x* [Tarski, 1983a, 195].

This is hardly something of which Tarski's readers had "intuitive" knowledge beforehand and it can't be found discussed in works on the "theory of knowledge" prior to Tarski's work. It also, famously, ties the defined term to the specific features of Tarski's "language of the calculus of classes"; yet surely it was no part of the concept of truth of which readers could have had intuitive knowledge that sentences of a certain first-order theory of the subset relation are true as a degenerate case of the satisfaction relation. Whatever this definition is for, it can't in any familiar sense be an "analysis" of the concept of truth. So in what sense, then, is the work concerned with "the concept of truth in formalized languages"?

Introducing the definition of truth is not the only topic that moves Tarski to insist that the "intuitive meaning" or "concept expressed by" some *definiendum* is in some way perfectly clear as a prelude to giving a definition that seems a far cry from something that analyzes this supposedly clear concept. Tarski's 1930 paper on definability, in the sense of the definition of a set by a predicate, rings a similar theme:

> ... we can try to define the sense of the following phrase: 'A finite sequence of objects satisfies a given sentential function.' The successful accomplishment of this task raises difficulties which are greater than would appear at first sight. However, in whatever form and to whatever degree we do succeed in solving this problem, the intuitive meaning of the above phrase seems clear and unambiguous (1983, 117).

This theme of rendering intuitions precise, related to use of the phrase "intuitive meaning" also occurs earlier in the same article:

> The problem set in this article belongs in principle to the type of problems which frequently occur in the course of mathematical investigations. Our interest is directed towards a term of which we can give an account that is more or less precise in its intuitive content (*contenu intuitif*), but the significance (*signification*) of which has not at present been rigorously established, at least in mathematics. We then seek to construct a definition of this term which, while satisfying the requirements of methodological rigour, will also render adequately and precisely (*avec justesse et précision*) the actual meaning (*signification "trouvée"*) of the term ...
>
> I shall begin by presenting to the reader the content (*contenu*) of this term, especially as it is now understood in metamathematics. The remarks I am about to make are not at all necessary for the considerations that will follow—any more than empirical knowledge of lines and surfaces is necessary for a mathematical theory of geometry. These remarks will allow us to grasp more easily the constructions explained in the following section and, above all, to judge whether or not they convey the actual meaning (*signification "trouvée"*) of the term [Tarski, 1983f, 112] ([Tarski, 1986g, 519]).

The passage has many noteworthy features. In the first paragraph we learn that "definable set of real numbers" has an intuitive content, a "found meaning" (Woodger renders *"trouvée"* as "actual" and unhelpfully drops the scare-quotes) which might be "rigorously established" in mathematics but so far has not been. The picture appears to be one in which a meaning grasped outside of a rigorous account has to be *imported into* or *expressed* in a formalism. On this same topic, the subsequent sentence tells us that the task at hand is to "render adequately and precisely" this "found" meaning, again suggesting that the issue at hand is how to ensure that a formalism adequately captures a certain known meaning. However, as with CTFL, the definition Tarski actually gives [Tarski, 1983f, 128] doesn't seem to have anything to do with the intuitive meaning of the defined term.

The quotation tells us that the formal theory that is constructed in order to express intuitions is independent of them in that it *can* be understood even by someone who lacks the intuitions or lacks knowledge that the theory is supposed to express. It is an "external" purpose of a formal treatment of something to capture an intuitive content; the resulting bit of mathematics can be understood independently, but our purpose in setting it out is to express our intuitions. This also gives us a bit more information about how Tarski is seeing the relationship between an "intuitive content" and its formal expression. Tarski announces that he will give us a "presentation" of the content of "definable set". Presumably such a presentation is something like a conceptual analysis—Tarski will make some claims about the "found" meaning of the term. This analysis will then guide the formal project of definition. This structure is the same as the one we have already seen in CTFL: there is a concept that is supposed to be somehow encompassed by a definition, the definition itself, and between them there is a conceptual analysis of a more traditional sort that informs or guides the setting out of the definition.

So the intuitive notions are important, but the formal system and the intuitive notions are relatively independent. On the one hand, as noted, the definition and the formal treatment of which it is a part can be understood even by someone who lacks the intuitive notion that the formal treatment captures. On the other, intuitive content is independent enough of its "rigorous expression" that it exists and can be grasped without it. Although "Every reader possesses in greater or less degree intuitive knowledge of the concept of truth" nevertheless:

> In §1 colloquial language is the object of our investigations. The results are entirely negative. With respect to this language not only does the definition of truth seem to be impossible, but even the consistent use of this concept in conformity with the laws of logic [Tarski, 1983a, 153].

It is significant here that Tarski says not what he is often presented as saying [Soames, 1999, 51], that the ordinary concept of truth itself is somehow "inconsistent" or otherwise incoherent, but rather that it cannot be "used in conformity with the laws of logic" in application to "everyday" language. Since a definition of the sort Tarski seeks cannot be provided for colloquial language, but "every reader" possesses "intuitive knowledge" of the concept of truth, grasp of the concept is independent of the sort of definition Tarski wants to provide.

Clearly, the relation between Tarski's definitions of semantic terms and "conceptual analysis" of the usual sort isn't, as one might expect, that the definition simply captures an analysis conceived of as something like a set of claims that anyone who grasps the concept must accept.

Summing up what we have seen so far, on Tarski's view concepts are independent of the terms that express them and are "intuitively" grasped. Definitions within rigorous mathematical treatments are supposed to aid the expression of intuitively grasped concepts or render the "found meaning" of ordinary terms. Between a concept and a definition of a term that expresses it within a particular theory lies something like conceptual analysis traditionally conceived. Finally, concepts and rigorous definitions are independent enough that each can be grasped without the other; one can understand a formalism intended to express a concept without grasping the concept or the fact that the formalism is supposed to capture it, and one can grasp the concept without access to any definition of Tarski's sort. Lurking in the background of Tarski's project, then, appears to be a rather articulate view of concepts, meanings, definitions, formalism, ordinary language, and related topics—a set of views in the philosophy of logic, language and to some extent mind. In order to understand Tarski's work we need to know what these views were.

1.1.2 Tarski, Leśniewski and Intuitionistic Formalism

In the introduction to an early article on the consequence relation Tarski expresses allegiance to a doctrine he attributes to Leśniewski:

> In conclusion it should be noted that no particular philosophical standpoint regarding the foundations of mathematics is presupposed in the present work. Only incidentally, therefore I may mention that my personal attitude towards this question agrees in principle with that which has found emphatic expression in the writings of S. Leśniewski and which I would call *Intuitionistic Formalism* [Tarski, 1983c, 62].

The main contention of these first two chapters will be that the aspects of Tarski's view just enumerated are aspects of Intuitionistic Formalism as he conceives of it. Note that this passage shares something with the passage from "On Definable Sets" that we discussed in the previous subsection: Tarski insists that certain more "philosophical" views are not strictly relevant to our understanding of certain formal work, yet finds the philosophical views worth mentioning anyway. Unfortunately, the result of this is that Tarski tells us almost nothing about what the "Intuitionistic Formalist" believes.

We thus turn for illumination to the passage from Leśniewski's *Grundzüge eines neuen Systems der Grundlagen der Mathematik* that Tarski cites. It bears mentioning that the cited passage is preceded by 15 pages of formal syntax in Leśniewski's meticulous version leading up to the statement of the "directives" (rules of inference and definition) for SS5, one of the many versions of "Protothetic", Leśniewski's system of sentential logic, discussed in the article:

Perhaps I should add that for many months I spent a great deal of time working systematically towards the formulation of these systems of Protothetic by means of a clear formlation of their directives using the various auxiliary terms whose meanings (*Bedeutungen*) I have fixed in the terminological explanations given above. Having no predilection for various 'mathematical games' that consist in writing out according to one or another conventional rule various more or less picturesque formulae which need not be meaningful or even—as some of the 'mathematical gamers' might prefer—which should necessarily be meaningless, I would not have taken the trouble to systematize and to often check quite scrupulously the directives of my system, had I not imputed to its theses a certain specific and completely determined sense (*Sinn*), in virtue of which its axioms, definitions, and final directives (as encoded for SS5), have for me an irresistible intuitive validity (*intuitive Geltung*). I see no contradiction, therefore, in saying that I advocate a rather radical 'formalism' in the construction of my system even though I am an obdurate 'intuitionist'. Having endeavored to express some of my thoughts on various particular topics by representing them as a series of propositions meaningful (*sinnvoller Sätze*) in various deductive theories, and to derive one proposition from others in a way that would harmonize with the way I finally considered intuitively binding (*welche Ich "intuitiv" als für mich bindend betrachte*), I know no method more effective for acquainting the reader with my logical intuitions (*"logischen Intuitionen"*) than the method of formalizing any deductive theory to be set forth. By no means do theories under the influence of such a formalization cease to consist of genuinely meaningful propositions which for me are intuitively valid [Leśniewski, 1992e, 487] [Leśniewski, 1929, 78].

Note, first, that Leśniewski speaks of *Sätze* being meaningful "in" a deductive theory. As we will see, this idea of a sentence being meaningful "in a theory" is far from idle. Second, like Tarski's translator Woodger, Leśniewski's translator O'Neil has a bad habit of dropping scare-quotes. Leśniewski finds the appeal to intuition problematic, but also finds that he has no other way to say what he wants to say.

Leśniewski here contrasts a certain parody of formalism (a caricature he associated with Hilbert; see e.g. [Leśniewski, 1992h, 195]) with his own approach which, though it involves formalization, aims at the expression of "intuitively valid" thoughts. Two central elements of the conception expressed in the passage are the usage of "formulas" in accord with "conventional rules" and the expression of thoughts. Lesniewski insists that the point of the use of formulas in accord with rules is to express thoughts, and also that designing the rules so that they adequately express the thoughts is a painstaking labor. The "formulas" in a deductive system, then, are assigned two roles: they stand in relationships to other formulas in accord with conventional rules and, at least when things go well, they express thoughts. So far, then, this places Leśniewski squarely within the "expressive semantics" camp discussed in the introduction: aside from language–language relations, Leśniewski's account of what Intuitionistic Formalism is about here mentions only language–mind relations as the locus of linguistic significance. This is so far consonant with what we have found in Tarski. For Leśniewski a deductive science consists of symbols governed by rules or conventions, but the formulas, used in accord with the rules, are supposed to express thoughts—just as for Tarski, on the one hand, we have something intuitive and clear that is supposed to be expressed within a deductive science, and on the other hand we have the deductive science itself, which may be quite fastidiously constructed, and constructed in a way that seems, at least to the uninitiated, to lack the overt clarity that the intuitive concepts already have.

One thing to notice in the passage is the conception of inference it offers. Leśniewski says that the "intuitionistic" aspect of his "formalism" consists in part in the fact that he "deduces one sentence from other sentences in a manner that harmonizes with the inferences he considers intuitively binding". This means that a central plank of Intuitionistic Formalism, in addition to the intuitive "validity" (*Geltung*—this being a common term at the time for the status of theorems among those, e.g. [Carnap, 1929], who had objections to simply calling such things "true") of axioms and definition—is the intuitive validity (in the usual sense) of the inferences among the sentences of a deductive theory. Leśniewski demands that derivations be "intuitively binding". Ruled out here, then, is a conception of validity or logical consequence that doesn't involve such an epistemic aspect. A central aspect of the doctrine to which Tarski swears fealty in 1930, then, is an epistemic conception of consequence.

As we will see in subsequent chapters, Tarski introduces his treatment of semantic concepts as a contribution to Intuitionistic Formalism. But since the introduction of the semantic notions takes place *within* the framework of Intuitionistic Formalism, it takes Tarski a good deal of time—all the way until 1935—to venture to replace Intuitionistic Formalism's epistemic conception of inference with a generalist conception based in his semantics.

1.1.3 Formalism

Since Intuitionistic Formalism was supposedly a kind of formalism, we should remind ourselves of what the relevant authors to whom Leśniewski and Tarski allude or refer had to say. Turning to Hilbert himself, we find remarks like this one, from the programmatic "Mathematical Problems" of 1900, to the effect that the meaning of an expression is somehow determined by the axioms in which it figures:[1]

> When we are engaged in investigating the foundations of a science, we must set up a system of axioms which contains an exact and complete description of the relations subsisting between the elementary ideas of that science. The axioms so set up are at the same time the definitions of those elementary ideas; and no statement within the realm of the science whose foundations we are testing is held to be correct unless it can be derived from those axioms by means of a finite number of logical steps. [Ewald, 1996, 1104]

Detlefsen, in his survey, characterizes the formalist position as follows:

> The mathematician, on this view, is free to stipulate of a concept she's introducing that it have exclusively the properties provided for by the axioms she uses to introduce it. There is no content belonging to a concept introduced in this way except that which is provided for by the introducing axioms. The axioms used to introduce a concept thus guaranteedly take on a kind of completeness—what might be called *constitutive completeness* because, together, they *constitute* the 'content' (i.e. the role in reasoning) of the concept.
>
> Concepts are thus identified with the *roles* they occupy in mathematical thinking (*denken*). They do not have to have an intuitive content to be significant. Hilbert thus believed that the inferential roles of concepts are determined not by contents given prior to the axioms which introduce them, but by those introducing axioms themselves. Indeed, all reasoning concerning a concept is *restricted* to that which is provided for by its introducing axioms [Detlefsen, 2008, 294–5].

What one cannot find in this rendition of formalism is anything like Leśniewski's insistence on the expression of "intuitively valid" thoughts and the like; indeed, as characterized by Detlefsen, the whole point of the view is to be independent of such things.[2] One can see this view in action in a later work by Carnap that came to be important to Tarski, *Logical Syntax*:

Up to now, in constructing a language, the procedure has usually been, first to assign a meaning to the fundamental mathematico-logical symbols, and then to consider what sentences and inferences are seen to be logically correct in accordance with this meaning. Since the assignment of the meaning is expressed in words, and is, in consequence, inexact, no conclusion arrived at in this way can very well be otherwise than inexact and ambiguous. The connection will only become clear when approached from the opposite direction: let any postulates and any rules of inference be chosen arbitrarily; then this choice, whatever it may be, will determine what meaning is to be assigned to the fundamental logical symbols [Carnap, 2002, xv].

This is quite the opposite of Leśniewski's view; rather than the intuitive content that lies outside of the deductive system being the important thing, the difficulty of determining it rigorously is for Carnap in the passage precisely a reason to get rid of it. However, we should remember here that in passages we have already looked at Tarski says something similar: the intuitive content of his defined terms is of interest, but his definitions and the theories to which they are added can also be understood independently of it. Nevertheless, for Tarski and Leśniewski extra-systemic intuitive meaning drives the construction of a deductive theory.

1.2   Leśniewski

1.2.1   Leśniewski's early work

When we turn to Leśniewski's early (1911–1914) papers, we find some much more articulate comments on a view which is in agreement with the remarks in the passage Tarski cites. An appeal to the early works is problematic, however, since Leśniewski later repudiated them [Leśniewski, 1992h, 197–8]. Nevertheless, to repudiate one's juvenilia is not to give up every view expressed in it, and one thing that remained in the later period, I will argue, is Intuitionistic Formalism.[3] After looking at the evidence from the early work I will confirm this view, and bolster my attribution of it to Tarski, in two ways: first by chasing down its traces in Leśniewski's later papers, and second by finding exactly the same view in the main work of Tarski's other mentor, Tadeusz Kotarbiński, published in 1929—the year in which Leśniewski published the passage Tarski cites, and in which Tarski did the bulk of the work on his strategy for defining truth [Hodges, 2008, 120–5].

The most illustrative passage from Leśniewski's early papers comes from his 1911 discussion of "existential propositions", the topic of his dissertation [Woleński, 1989, 8]. Having argued for a series of rather radical claims (e.g. that all positive existentials are analytic) he suggests that his views appear more paradoxical than they are because people are confused about how to express thoughts clearly: it is simply a confusion, for instance, to think that "people exist" is the right way to put the thought often expressed by it; better for this purpose is "some beings are people" [Leśniewski, 1992b, 15]. Since saying this commits him to views about the right way to express thoughts, Leśniewski continues:

> I have discussed, in this section, the adequate and inadequate representation of various contents by means of various propositions. An obvious question might therefore arise, what is my criterion of adequacy or inadequacy of the representation of given contents in a proposition...
>
> The symbolic functions (*funkcje symboliczne*) of complex linguistic constructions, e.g., propositions, depend on the symbolic functions of the elements of these constructions, that is individual words, and on their mutual relationship. In the unplanned process of development of language, the symbolic function of propositions can depend in some particular cases on identical symbolic functions and on identical relationship between specific words—in quite different ways. The planned construction of complex linguistic forms cannot, for representing various contents (*treści*) in the system of theoretical propositions, be confined within the possible results of the unplanned evolution of language. Such construction calls for the formation of certain general conventional-normative schemata (*schematów konwencjonalno-normatywnych*) to embody the dependence of the symbolic function of propositions on the symbolic functions of their elements, and on the mutual relationship between these elements. To ascertain whether the given content has been represented adequately or inadequately (*adekwatności lub nicadekwatności*) in a proposition, one has to analyze individually how the speaker's representational intentions (*intencji symbolizatorskich mówiką cego*) relate to the above-mentioned schemas. These schemas should indicate in what way the symbolic function of a proposition should be conditioned by the symbolic functions of the particular words and by their mutual relationship. To express this dependence I have adopted the following normative schema: every proposition is to represent (*symbolizować*) the possessing, by the object denoted by the subject, of the properties connoted by the predicate. My assertion that a

given proposition does not adequately represent the given contents is always based on the fact that the possession by an object represented by the subject of a proposition of certain properties connoted by the predicate does not imply the same content that the speaker intended to represent ... The semiotic analysis of the adequacy or inadequacy of certain propositions in relation to the contents which they represent is then ultimately based on a phenomenological analysis of the speaker's representational intentions [Leśniewski, 1992b, 16–17] [Leśniewski, 1911, 342–3].[4]

Here as in 1929 the function of language is to represent the mental "contents" speakers intend to convey. Leśniewski refers in the passage to "phenomenological analysis" of the speaker's representational intentions and earlier in the paper to Husserl [Leśniewski, 1992b, 6] and, following Twardowski (e.g. [Twardowski, 1977, 8–10]),[5] conceives of intuitive content throughout the paper in terms of the notion of connotation from traditional logic, with reference to Mill.[6] The main point for us is that the adequacy of conventions here is a matter of properly answering to a speaker's "referential intentions". The point of language is to express thoughts, but natural language does this poorly and the purpose of deductive sciences is to do it well. A system of language needs to be rigorously compositional in the sense that the content expressed by a non-lexical expression must always be uniquely determined by the contents associated with its parts and by the way they are assembled. The "symbolic function" of expressions in such a system is determined by "conventional–normative" schemas: rules associating thoughts and thought-constituents that are to be expressed by the expressions of the system with these expressions.

This sounds good, but it raises a problem for Leśniewski, one that neither he nor Tarski ever solved or ever really even addressed: the proposed account of compositionality for contents didn't exist.[7] There is here a simple picture of contents of primitive terms, construed as sets of properties in accord with traditional logical doctrines about connotation, and then hand-waving about how the contents of complex expressions have to be determined by the contents of simple ones plus their manner of composition. One can see how this might go in a few toy cases (e.g. "red ball"), but it is unclear what the story was supposed to be for anything more complicated. This weakness in Intuitionistic Formalism was never repaired. When Tarski replaced intutionistic formalism with his semantics, one result was that an account of meaning with no story about semantic compositionality was replaced with another that had one, but that there was this difference seems not to have been something that motivated Tarski.

Clearly the view Leśniewski adheres to in 1911 is retained in the passage Tarski cites and in the material surrounding it, despite Leśniewski's repudiation of the rather outdated and idiosyncratic views of logic and meaning present in the article. "Intuitionistic Formalism" is then the view that the point of formal systems is the clear expression of thought, and that such systems are to be judged "adequate" or not based on how well their expressions, used in accord with "conventional–normative schemata", systematically accord with the representational intentions of their "speakers".

One can find elements of this view throughout Leśniewski's work. Let us trace the appearance of its main elements—the conception of meaning in terms of representational intentions with respect to intuitively grasped thoughts, the conception of the "conventional–normative schemata" governing the use of language in terms of this, and the conception of intuitive thought in terms of the traditional doctrine of connotation and denotation—through his work.[8] We will first look at expressions of the view in the early papers, and then find its appearance in the mature work.

The content expressed by a term as conceived by Leśniewski is a connotation. We can see this in his account of synonymy at [Leśniewski, 1992a, 21–22]: two subject-predicate propositions are synonymous when both their subjects and their predicates have the same connotation. We can also see it in his remarks on non-denoting expressions that are nevertheless meaningful: "'hippocentaur' does not denote any object … This does not mean, of course, that this word does not connote anything—on the contrary, it can connote some quite strictly defined properties … and thus it can possess a strictly defined 'meaning'" [Leśniewski, 1992a, 32]. At one point in the early work Leśniewski writes that the full form of the definition of an expression is illustrated by "I employ the expression 'W' in such a way that if it denotes anything at all, it denotes only those objects which possess the properties $c_1, c_2, c_3, \ldots, c_n$" [Leśniewski, 1992c, 66]. Note here the conception of a definition as a convention that accords with a speaker's referential intentions.

Leśniewski is quite serious in thinking of language as functioning entirely to express thoughts rather than, for instance, to shape or determine them; witness this passage from a subsequent article in which the rules of language are compared to railway signals:

I have more than once pointed out that a system of linguistic symbols, just as any other system of symbols, e.g., the system of railway signals, requires the existence of certain rules for constructing the symbols and keys for reading them. I have repeatedly stressed that the functions of various complex linguistic structures, e.g., those of propositions, should depend, in a correctly constructed precise language, upon the functions or the order of particular words—on the bases of certain pattens determined by general normative conventions the knowledge of which permits the correct symbolization of an object in a given language or the decoding of a symbol for a given language [Leśniewski, 1992c, 56].

Note the use, here and elsewhere, of "decode" (*odcyfrowanie* [Leśniewski, 1913, 324]) for what the reader does in understanding a symbol in accord with conventional-normative schemata. This passage is followed by the statement of four "Conventions" in accord with which Leśniewski will use certain expressions. These mostly determine the role of negation and allow the construction of Leśniewski's argument, such as it is, for the principle of non-contradiction: given the four conventions, all contradictions are false. The main point for us in this is simply that a "Convention" is a rule in accord with which one intends to use a symbol: a point that will be relevant when we get to Tarski's "Convention" T.

The conventional–normative schemata of a deductive science are, then, set up by Leśniewski to express his thoughts. Leśniewski's account of the latter, however, accords them no particular justificatory status. Consider these remarks from an early discussion of his mereological conception of set theory:

The psychic 'sources' of my axioms are my intuitions, which simply means, that I believe in the truth of my axioms, but I am unable to say why I believe, since I am not acquainted with the theory of causality. My axioms do not have a logical 'source', which simply means that these axioms do not have proofs within my system [Leśniewski, 1992d, 130].

Appeals to intuition by Leśniewski are not to be understood as having any justificatory force; the whole point of the axioms is simply to express clearly the author's thoughts.[9] This will be important when we turn later to the question of whether the Intuitionistic Formalist followed Hilbert in taking consistency to be sufficient for truth.

With these basic views on meaning in mind, we can turn to Leśniewski's conception of "science" as embodied in a deductive theory:

Science is, then, a system of linguistic symbols. To construct and to understand linguistic symbols, it is necessary, as in the case of any other symbols, to have certain rules of constructing these symbols and certain keys to decipher them. The rules on which the construction of linguistic symbols are based and the keys which decipher them are, on the one hand, precise definitions of various expressions and, on the other, different kinds of conventions concerning linguistic symbols ... Linguistic conventions are, therefore, the necessary condition of any scientific procedure and the indispensible key to understanding science [Leśniewski, 1992a, 35].

As Leśniewski mentions elsewhere, such conventions can amount to definitions or not [Leśniewski, 1992c, 75–6]: either way they are rules that determine the intuitive meanings of symbols.

A passage from the transitional "Foundations of the General Theory of Sets" of 1916 already states the Intuitionistic Formalist view well:

On the question concerning the way of using expressions, I mention that of the mathematical terms which I use, the expression 'part' is the only one I do not define, supposing that this term will not cause misunderstandings—considering that its intuitive character acquires considerable clarity in the light of Axioms I and II [Leśniewski, 1992d, 131].

Axioms I and II state that parthood is anti-symmetric and transitive, respectively. Unlike the pure formalists, Leśniewski holds that the intuitive content of the primitive expression matters. One function of the axioms is to express this content sufficiently to ward off misunderstandings.

1.2.2 Leśniewski's later work

After Leśniewski's relatively silent period of teaching and working diligently at Warsaw he resumed publication with "On the Foundations of Mathematics" (1927–1931) and repudiated the early work. His views on the point of setting out a deductive theory receded into the background, but traces of them can still be found in remarks throughout the remainder of his writings, including the passage Tarski cites. One thing that these later remarks make clear is that though Leśniewski made some reference to Brentano and Husserl in his early discussions of "referential intentions" and the like, no heavy theory of "mental acts" lies behind his views; his interest is really a practical one in being understood to have said exactly what he intended to say.

Leśniewski retakes the philosophical stage after his absence with a trenchant criticism of known systems of mathematical logic, criticizing the major contributions of the period—Frege's proposed modification to his own system in the appendix to the *Grundgesetze*, *Principia*, Zermelo and others—for their lack of intuitive backing. Consider this passage:

> The problem of the 'antinomies', under the powerful influence of the researches of Bertrand Russell, has become the central problem in the intellectual endeavors of a number of prominent mathematicians. These endeavors were often significantly remote from the historical, intuitive basis from which the 'antinomies' developed. They encouraged the disappearance of the feeling for the distinction between the mathematical sciences, conceived of as deductive theories, which serve to capture various realities of the world in the most exact laws possible, and such non-contradictory deductive systems, which indeed ensure the possibility of obtaining, on their basis, an abundance of ever new theorems, but which simultaneously distinguish themselves by the lack of any connection with reality of any intuitive, scientific value ...
>
> The question, whether the Frege system changed as indicated above or also the '*Mengenlehre*' of Zermelo will ever lead to a contradiction, is completely immaterial from the point of view of an intellectual anxiety, directed resolutely towards reality, and which stems from an irresistible, intuitive necessity of belief in the 'truth' of certain assumptions and in the 'correctness' of certain inferences, which in combination with those assumptions, lead to contradiction. From this point of view the only real 'solution' to the 'antinomies' is the method of an intuitive undermining of the combination of inferences or assumptions which contribute to the contradiction. A non-intuitive mathematics contains no effective remedy for any malady of the intuition. [Leśniewski, 1992h, 177–8].

As distinct from a pure formalist, or the Carnap of *Logical Syntax*, Leśniewski sees difference between a deductive theory that expresses intuitive thought about reality and a merely consistent system as being of great import.[10] The passage is followed by a long series of complaints about the inadequacies of Russell and Whitehead's attempts to explain the meanings of the expressions of the language of *Principia*. The only predecessor for whom Leśniewski expresses any approval is Frege [Leśniewski, 1992h, 195 note 18], for his attention to the "meaningfulness of mathematical propositions".[11] Our interest in the passage concerns its continuity with what we have seen: the point of a deductive science is clearly to express thoughts accepted as intuitively true; a mere formalism alone cannot solve any problems.

Commenting on his turn to "a systematic use of symbolic language instead of a colloquial one in the year 1920", Lesniewski makes clear that his interest was simply in warding off misunderstanding:

> The symbolism I adopted, based on formulas constructed by 'mathematical logicians', I used as a tool which is technically much simpler than the colloquial language and at the same time less prone than that language to lead to misunderstandings in the formulation of ideas. In attempting to translate the theses of my 'general theory of sets' as scrupulously as possible from the colloquial language into my new 'symbolic language', I nevertheless constructed the proofs of the theorems of that theory in an intuitive way without in any way basing those proofs upon some clearly codified system of 'mathematical logic', a procedure I had used up to then. The change to a 'symbolic' way of writing, which constituted a far-reaching revolution in my scientific life in the field of symbolic technique, was not accompanied by any far-reaching parallel events in the domain of my 'logical' views [Leśniewski, 1992h, 365–6].

Later Leśniewski did become nearly obsessed with the construction of formal rules that captured these "intuitively binding" inferences, in the "directives" of his systems. What we can notice here is that Leśniewski wanted to express his thoughts with the minimum possible risk of misunderstanding. His thoughts, and his inferences among them, took place in his mind; what he sought was a medium in which they could be expressed. Note that the conception of inference here is entirely epistemic: proofs need to be "intuitively binding"; the later introduction of explicitly formulated "directives" for adding lines to a proof simply makes for an added degree of exactness in expressing such inferences.

Since the point of the symbolism was to ward off mistakes on the part of the reader as to what ideas Leśniewski was trying to convey as much as possible, an overriding goal of the symbolism was thereby to reduce those aspects of the symbolism that weren't specified in terms of the syntactic forms of expressions to a minimum. Thus began the obsession among Leśniewski and his students with reducing the number of axioms and primitives in a deductive theory. This project, which can look like meticulousness for its own sake, was well-motivated on Intuitionistic Formalist grounds: the fewer the primitives, the fewer the signs whose meanings had to be explained in "colloquial" language; the fewer the axioms, the fewer the claims upon which agreement would need to be reached before one could get down to the business of deriving intuitively valid claims.

One aspect of Leśniewski's vision here is that axioms and definitions mattered less to the intuitive interpretation thereby forced upon his primitives than did certain theorems. That is, Leśniewski, in order to capture intuitive meaning, wanted certain theses to be theorems of his system so as to guarantee its expression of intuitive content, but he was more open about the formulation of axioms and definitions. Here is how he comments on the formalization of his "Ontology":

> I wished to take one step further and, using the 'singular' propositions of the type 'Aεb', to base all my deliberations on some clearly formulated axiomatization which would harmonize with my theoretical practice in this domain at the time. In relation to such an axiomatization I required that no 'constant' terms should appear in it with the exception of the expression 'ε' in propositions of the type 'Aεb' and terms which appear in the 'deductive theory' ...
>
> Feeling the need of a deductive theory axiomatized in the way described, I nevertheless could not find a ready theory of that sort in the work of either the 'traditional logicians' or the 'mathematical logicians' ... I one day formulated ... a number of theses which would be valid on the basis of the theory mentioned. I cannot recall which theses I mentioned then—after so many years—but I can state with complete confidence ... that at that time I would not have accepted as sufficient for my theoretical purposes, as described above any axiomatization which did not guarantee the possibility of deriving from it—while permitting, of course, operations with definitions which harmonized with my way of using the corresponding expressions in my scientific practice,—'symbolic' equivalents of theses stating respectively that
>
> (1) some *a* is *b* when and only when, for some *X*, (*X* is *a*, and *X* is *b*) ...
>
> (6) *A* is *a* when and only when, (each *A* is *a*, and at most one object is *A*).

From among the different theses, whose 'symbolic' equivalents were valid in the deductive theory which I wished to build, I have selected here theses (1)–(6), as they appear to me a convenient intuitive transition to other deliberations: thesis (6), or any other thesis more or less similar to (6), has played a far-reaching role at the creation of the aforesaid theory, whose sole Axiom I devised by analysis of that thesis; theses (1)–(5), selected to avoid intuitive difficulties for the reader, throw some light (directly or indirectly) on the way in which I used the expressions appearing in the thesis (6) [Leśniewski, 1992h, 367–8].

Here, quite clearly, the role of certain theses in the system is to secure agreement as to the meaning of the primitive terms of the system, and the desired axiomatization is held to the standard that they be implied. By contrast, the axiom itself is simply required to imply (1)–(6) by the directives; it need not be intuitive in its own right.

That said, implication of (1)–(6) of course didn't make the system immune to misunderstanding. Leśniewski comments at one point on the topic of securing agreement about the primitive term of his "Ontology":

> It often happened that someone with whom I spoke did not real-ize the sense in which I was using the sign 'ε' in my ontology, and experienced interpretational difficulties in attempting to relate the given sign semantically to one or other signs handed down by the tradition of 'mathematical logic' and 'the theory of sets'. In the trou-blesome theoretical situations in which I found myself due to this state of affairs, I shared the misfortune of all who are forced by cir-cumstances to convey in 'their own words' the sense of the various primitive terms of deductive theories which they were constructing [Leśniewski, 1992h, 374–5].

The remark is then followed by a discussion of various "colloquial" formulations that might help someone to cotton on to Leśniewski's sense. Notice again the contrast here with Hilbert and especially with Carnap, whom we have seen in *Logical Syntax* taking this very prob-lem to be the reason that one ought to ignore anything that isn't determined by the structure of a deductive theory. For Leśniewski, by contrast, the colloquial explanation, though problematic, was of central importance.

One change in the shift from the early to the late view that is likely asso-ciated with the move to consideration of formalized languages is that the early work is concerned primarily with sub-sentential conventions, while mention of these is absent in the later passage. On Leśniewski's mature view conventions that determine the intuitive thoughts expressed by sentences are conceived of as the "directives" governing a system, rules for adding new theses to it. The determination of sub-sentential mean-ings expressed is then an indirect matter of the role of a sign as established by the theorems in which it appears, and it is this determination that comes to be the central issue, as with the example of "ε" above. This shift de-emphasizes the earlier demand that the account be composi-tional and Leśniewski in the later work doesn't comment on that earlier requirement. As we will see in § 2.2.1, in Tarski's hands the Intuitionistic Formalist conception of intuitive meaning is not compositional.

The picture that emerges, then, is a rather simple one. Leśniewski has ideas and thoughts. There is no question for him as to exactly what they are, but he wants to convey them to another. For this an adequate symbolism is a better tool than natural language and so the thoughts are best expressed in the form of a formalized deductive theory governed by rules for the formation of expressions and for the conditions under which one sentence can be asserted on the basis of others. Since one point of the exercise is to rule out possibilities of misunderstanding, axioms and primitives are to be kept to a minimum and, since the other point of the exercise is to state (what Leśniewski is convinced are) truths, what matters is that the axioms seem true—recall here Leśniewski's entirely non-epistemic, psychologistic notion of "intuition"—and that the inferences from them to theorems seem valid.[12] Since relations instituted in a deductive science by conventions are concerned with sentences, a primary issue is the way in which sentences held to be true constrain the possible interpretation of the primitives. This is Intuitionistic Formalism.[13]

On Leśniewski's conception the point of a convention—and, since definitions are a kind of convention for Leśniewski, the point of a definition—is to determine, jointly with other conventions, what intuitive concept a term expresses, not by directly analyzing this content, but by implying things, in conjunction with other theses, and in accord with the directives, that constrain the interpretation of it. Definitions were, for Leśniewski, simply further axioms that contributed to constrain the concept expressed by a term; they differed from other axioms in showing also how to eliminate the term from theorems of the system. Leśniewski credits Tarski's work, published as [Tarski, 1983j], for allowing him fully to realize this conception of definitions for Protothetic, his system of sentential logic [Leśniewski, 1992e, 418–19], [Leśniewski, 1929, 11]. Thus there is room in Leśniewski's conception for definitions that do in fact, when added to a theory, determine that a term expresses a certain intuitive concept without somehow themselves being, or being expressions of, an analysis of the concept. What matters is simply that that connotation be properly captured in the system as a whole. What this anticipates is precisely the mismatch we have already examined between Tarski's definitions and what one would expect of traditional philosophical analyses of concepts, as well as the relationship between a definition of truth, the T-sentences and Convention T.

1.3   Kotarbiński

So far we have found a rather clear view expressed in Leśniewski's 1911–14 work, and then general statements consistent with it in later work. Since Tarski only began his studies in 1918, we need to ward off the suggestion that perhaps the view wasn't really around for him to inherit. However, turning to Kotarbiński [Kotarbiński, 1966], we find the same view as we have discerned in Leśniewski's early work. Tarski notes at the outset in CTFL that "in writing the present article I have repeatedly consulted this book and in many points adhered to the terminology there suggested" [Tarski, 1983a, 153]. To the extent that Tarski has views in the philosophy of language we can expect that they come from Kotarbiński, and since Kotarbiński's views are quite consistent with what we find in Leśniewski, we can be confident that the view sketched so far is in fact Tarski's "Intuitionistic Formalism".[14]

Kotarbiński's basic conception of meaning includes elements of intention and convention:

> That sentence S is a (direct) statement of John's thought as to content, means: John thought that p ("p" stands for the whole sentence) and uttered S so that the listener might recognize that John thought that p. Obviously, in such cases there is no obstacle to a given statement, which is not a direct statement of any person's thought as to content, being usable as such a statement in conformity with the laws of a given language, that is, expressing that thought as to content indirectly [Kotarbiński, 1966, 4].

The distinction between direct and indirect statements is simple enough; a sentence states an experience directly when "it is spoken or written with the intention that a definite person, or any person whatever who perceives that phrase, and understands the language in question, should know from that phrase that the speaker experienced precisely that" [Kotarbiński, 1966, 3–4], while "a given phrase states a certain experience indirectly whenever, without stating it directly, such can, in view of linguistic usage or adopted conventions, be used in a given language as a direct statement of such experience" [Kotarbiński, 1966, 4].[15]

Kotarbiński's phrase "the laws of a given language" matches Leśniewski's talk of "conventional–normative schemas". Kotarbiński offers no particular analysis of convention or "laws of language", nothing like an account of the conditions under which a convention is instituted in a population in terms of the mental states of its constituent people in the style familiar to us from [Lewis, 1969], but it isn't a far stretch to determine what he has to be assuming. One can intend to use a sentence to express a particular thought; this is direct statement. It can also be the case that the laws or conventions of a given language determine that that a particular sentence expresses some thought, where this comes to its being the case that a person who had a certain thought and used the language *correctly* would use that sentence directly to express it. So the assumption has to be that "conventions" and "laws" of language are agreed-upon rules that relate thoughts to the sentences that ought to be used, in that language, to express them.[16] A convention then obtains in a population when its members do agree that the sentences ought to be used directly to express those thoughts. Once such a convention is in place, a phrase indirectly expresses on all occasions that thought which it would be used directly to express if it were used correctly by someone.[17] Consider here this remark on the more general phenomenon of "expressing", which is basically statement less the intention to communicate:

> The word 'help!', exclaimed by John at the moment of danger, expresses directly only the desire, experienced at a given moment, to obtain help. But indirectly, in view of the English language usage, it expresses also a similar desire which Peter would have expressed even if it happened that he was drowned without being able to call for help. What is more, the word 'fy', shaped by chance from the twists of a creeping plant, expresses in the English language indirectly (and, of course, unintentionally) the experiences of disgust of various persons, even if those persons did not intend to express them and could not speak English. [Kotarbiński, 1966, 3].

The last example makes clear that the "rules" are directed at words and signs conceived of as items with certain *shapes* or their acoustic analogues.[18] An important element, though, is the way in which conventions relate the form of an expression to its meaning. This link opens up the possibility of a purely formal study of meaning, provided that the conventions are given. The idea that conventions relate syntactic form to intuitive meaning will be the source of Tarski's ideas about metamathematics as the study of systems of meaningful sentences where their forms go proxy for their meanings [Tarski, 1983c, 62].

This account of direct and indirect statements, combined with the role of convention in indirect statement, opens the space for precisely the problem that exercised Leśniewski: under what conditions does a given sign indirectly express exactly that which one wants directly to express with it? The set-up calls for a program of good hygiene in designing conventions for the indirect expression of the thoughts one wants directly to express about a given topic. It also makes clear that something is taken to be unproblematic by Kotarbiński, just as it was by Leśniewski: thoughts are "experienced", and presumably one knows what one's experiences are; in this the members of the Lvov–Warsaw School are good Brentanists.[19] The project in setting up a language is to devise conventions that are clear and agreed upon under which sentences indirectly express exactly what one wants directly to express by them. This done, one isn't left to the happenstance of a particular communicative situation to get one's thoughts across; if all goes well, fellow parties to the conventions will take sentences indirectly to express exactly what one directly expresses by them. Notice also, for future reference, that the emerging picture leaves open the possibility that the conventions of a language might fail: poor design could leave it unsettled what a sentence indirectly expresses, or determine an indirect expression that fails to accord with one's intentions or they might associate, if taken seriously, every thought with every sentence.

Kotarbiński's notion of meaning is the obvious auxiliary to his conception of indirect expression:

> If we want to answer the question "What does a given sentence mean?" we cannot describe a person's experience and say that this is "what that sentence means".
>
> We must try other ways, and probably the simplest consists in answering by the phrase "means that". If, for instance, someone asks what the saying "John is transported with joy" means, we may answer that the saying means that John is transported with joy [Kotarbiński, 1966, 5].

This applies, however, only to sentences. Kotarbiński continues:

> Someone might ask, for instance, what the word "goliard" means, which he has found in a text referring to the late Middle Ages. It would be absurd to answer: "goliard" means that etc., since the word "goliard" is not a sentence and hence does not state anyone's mental experience. It is only with reference to a sentence that we can reasonably say it "means that ... " ... How then shall we answer reasonably the question concerning the word "goliard"? The best solution is perhaps as follows: the sentence "A person is a goliard" means that the person is an itinerant minstrel composing satirical Latin verse. Whoever is, and rightly so, dissatisfied with "a person" can use the letter x instead and will arrive at the formula: the statement "x is a goliard"

means that x is an itinerant minstrel composing satirical Latin verse [Kotarbiński, 1966, 6].

In all those cases when someone asks what a given phrase means, the proper answer is obtained by using the phrase "means that", either used directly after the word or phrase to which the question pertains, or somehow implied in accordance with the syntactic function of the word or phrase... An analytic definition of the phrase "means that" as applied to sentences can be built with the aid of the words "states as to content" (or "is a statement as to content") in the following manner: "The sentence S means that q" is the same as "The sentence S states as to content, directly or indirectly, a person's thought that q" [Kotarbiński, 1966, 7].

We characterize the meaning of a sub-sentential expression only indirectly, by schematically indicating what a sentence including it will mean, while sentence meaning is, in turn, directly analyzed in terms of the expression of thought. Kotarbiński allows here "direct or indirect", but indirect expression is more to the point: we characterize what a sub-sentential expression means, according to linguistic usage, by indicating schematically what would indirectly be expressed by a sentence incorporating it when used correctly.

This schematic mode of specifying meaning bears an obvious connection to Tarski's notion of satisfaction. It also makes clear that conventions governing sentences relate to sub-sentential expressions only indirectly. For Kotarbiński the basic notion of meaning is that a sentence means that p if and only if the conventions of its language determine that it is used correctly exactly to express the belief that p.

Now, strictly speaking, what we have here is a theory of how "means that" and "means" are used: Kotarbiński has told us how to formulate an ascription of meaning. But what is it that is ascribed? We get the answer in Kotarbiński's account of connotation:

It is time now to introduce a new technical term, which will facilitate a further analysis of the relation of denoting, especially in the case of general terms. This is "connotation". Now instead of saying that, for example, "x is ice" means that x is solid water, we may say that the term "ice" connotes the properties of wateriness and solidity...

The *properties jointly connoted* by a given term form together the intension of that term, or its sense, or its connotation; this must be contrasted with the *designata* which jointly form the *extension of that term*, or its denotation. We may thus say that every object which has all the properties jointly connoted by a given term with respect to its given meaning is a *designatum* of that term with respect to that meaning. In other words, every object which has the intension of a given term with respect to its given meaning belongs to the extension of that term with respect to that meaning [Kotarbiński, 1966, 11].

The picture here is the same as in Leśniewski: terms have intensions that are sets of properties and their extension is everything that has the properties; as with Leśniewski, the source of the view is [Twardowski, 1977, 8–10]. Note that only "terms"—roughly, subjects and predicates of all orders—have connotations, while all syntactically significant expressions (e.g. sentential connectives [Kotarbiński, 1966, 6–7]) have meanings.

Thought and language are brought together here in the doctrine that the meaning of a term, its connotation, is also the intension of the concept expressed by it:

In semantic terminology, the intension of a concept is understood as the connotation of the corresponding term, that is the properties jointly connoted by the term. More strictly, the intension of the concept of M is the same as the set of properties jointly connoted by the term "M" [Kotarbiński, 1966, 88].

This doctrine of intension is the precursor of Tarski's notion of the content of a concept in CTFL.

We can consider the possibility of a mismatch that the account allows. On the one hand, terms' meanings are supposed to be determined by the conventions for sentences. On the other, a term's meaning is supposed to be a certain connotation, the content of a sub-propositional mental state: grasping a concept. The question raised is: how do we make conventions for sentences expressing thoughts correctly determine the connotations that terms are intended to have?

Though Kotarbiński discusses denotation at some length, the important thing for our purposes is that denotation is within his conception a derivative feature that is irrelevant to the basic, psychologistic account of the significance of linguistic expressions and their role in communication. For Kotarbiński, following in the footsteps of Twardowski, though thought and language can be characterized in referential, semantic terms, these terms play no role in the basic account of meaning and communication in language. Compare here Twardowski in a characteristic passage:

The word 'sun,' Mill maintains, is the name of the sun and not the name of our presentation of the sun; yet he does not want to deny that the presentation alone, and not the thing, is recalled through the name or communicated to a listener. The task of a name is thus twofold: the name communicates to a listener the content of a presentation and, at the same time, it names an object. However, we said that we must discern, not just a twofold, but a threefold aspect of every presentation: the act, the content, and the object. And if a name really yields an accurate linguistic picture of the mental state of affairs which corresponds to it, then it must also show a correlate to the act of presentation ...

What, then, is the task of names? Obviously, it is to arouse in the listener a certain content of a presentation. Someone who utters a name intends to awaken in the listener the same mental content which appears in himself; when someone says: "sun, moon, and stars," he wants those who listen to think, just as he does, of the sun, the moon, and the stars ... In this manner, a name already fulfills two tasks. Firstly, it makes known that the user of the name is presented with something; it signifies the existence of a mental act in the speaker. Secondly, it awakens in the listener a certain mental content. It is this content which is the "meaning" of a name.

[note] We designate in every case as the meaning of an expression that mental content which it is the real task, the goal, of the name to arouse in the listener. [Twardowski, 1977, 8–9].

As for Kotarbiński, thoughts are about objects and names refer to them. But, also as for Kotarbiński, the understanding of linguistic significance and meaning is entirely in terms of the thoughts and thought-constituents associated with them.[20] A significant linguistic act is one that is intended to arouse thoughts and ideas in an audience.[21] One mark of this is that Leśniewski and Kotarbiński find no puzzle about the linguistic significance of empty names: they have connotations, and this is what is required for being meaningful. Denotation is optional.

For Kotarbiński the rules of language determine which sentences indirectly express which thoughts; they are also supposed to determine, derivatively, that certain sub-sentential elements express "parts" of thoughts in the sense that thoughts about a given object, property, etc. are expressed by sentences that have common components. The issue that comes to the fore for Tarski concerns the conditions under which the conventions that determine which sentences are assertible correctly zero in on the connotations that are supposed to be expressed by various sub-sentential expressions. The main result of our examination of Kotarbiński is a grasp on this role of connotation in

Kotarbiński's account. Connotation stands between intuitive concepts and the conventional roles of the terms that express them. The main thing that the formal role of a term is supposed to capture is its connotation. We can thus be on the lookout for something similar in Tarski.

As with Leśniewski, Kotarbinski's conception of consequence in a deductive theory is explicitly epistemic and derivational:

> A deductive system is a set of sentences divided into two subsets, such that every sentence which belongs to the second subset has been derived exclusively from sentences belonging to the first subset, by means of transformations in accordance with knowledge-forming rules alone [Kotarbiński, 1966, 242].

Kotarbiński also shares with Leśniewski the view that the axioms of a deductive system need not themselves display any kind of self-evidence [Kotarbiński, 1966, 246ff]. This is a conception of consequence and the role of a deductive system shared with Leśniewski, and it is one that we will find that Tarski accepts at first as well.

Thus we find the same view in Kotarbiński's 1929 work as we found in Leśniewski's works of nearly 20 years earlier. The basic notion of meaning is that of the thought that a speaker intends to convey to an audience. Language is held to be governed by a system of conventional rules that determine which thought-contents are expressed when expressions are used correctly, and the theorems of a deductive science are supposed to be "intuitively valid", that is, to seem true given the conventions. Sentences held true as theorems are then supposed to determine the connotations expressed by the expressions they involve, but the question can be asked of any given proposed deductive science whether the system of conventions governing it adequately answers to speakers' "referential intentions" by ruling out unintended assignments of connotation to its terms on the part of its speakers.

1.4 Tarski in context

1.4.1 The axiomatic method

So far we have looked at the philosophy of language of Leśniewski and Kotarbiński, but we have yet to pay sufficient attention to their views about "deductive sciences" (axiomatic theories) in general. In order to see how Intuitionistic Formalism pans out as a view about deductive sciences, we can consider, in contrast to the view we find in Carnap in Hilbert, a line of reflection on the "axiomatic method" that runs from

a paper of Padoa that Tarski cites with enthusiasm, through Kotarbiński and on to remarks of Tarski that persist as late as 1937's *"Sur la méthode deductive"* and the inclusion of the latter, with additions, in Tarski's logic textbook. This sort of view adds to the formalist one the idea that the point or function of a deductive theory, though it can be understood on its own in the formalist manner, is to express certain intuitively valid thoughts of its creator. This line of thought thus joins the account of language we have found in Leśniewski and Kotarbiński to the formal study of deductive theories and, in that way, allows us to see further what "Intuionistic Formalism" meant to Leśniewski and Tarski.
Padoa writes:

> The necessary point of departure for any *deductive theory* whatsoever is a *system* of *undefined symbols* and a *system* of *unproved propositions*. Besides, it is obvious that a deductive theory has no *practical significance* if its undefined symbols and its unproved propositions do not represent (or cannot represent) *ideas* and *facts*, respectively. Thus the *psychological* origin of a deductive theory is *empirical*; nevertheless, its *logical* point of departure can be considered a *matter of convention*.
>
> Indeed, during the period of elaboration of any deductive theory we choose the *ideas* to be represented by the undefined symbols and the *facts* to be stated by the unproved propositions; but, when we begin to *formulate* the theory, we can imagine that the undefined symbols are *completely devoid of meaning* and that the unproved propositions (instead of stating *facts*, that is, *relations* between the *ideas* represented by the undefined symbols) are simply *conditions* imposed upon the undefined symbols.
>
> Then the *system* of *ideas* that we have initially chosen is simply one interpretation of the *system* of undefined symbols; but from the deductive point of view this interpretation can be ignored by the reader, who is free to replace it in his mind by *another interpretation* that satisfies the conditions stated by the *unproved propositions* [Padoa, 1967, 120–121].

In this quote we find a clear separation of the ideas "chosen" to be represented by undefined symbols and the facts "chosen" to be stated by the axioms from the basic formalist idea we have already found in Hilbert. What Padoa's conception adds is the idea that once the theory is constructed one need not know anything of the original ideas and thoughts its expressions and sentences were "chosen" to express, the idea that the theory is intended to allow a certain favored interpretation.[22] This is its "empirical" origin and the source of its "practical" significance.

Kotarbiński echoes the sentiment (not with citation of Padoa, though some later work of Padoa is cited by Kotarbiński) toward the end of a section on proving the independence and consistency of axioms through imposing varying interpretations on primitive terms, writing:

> It can be seen, however, both from the general characteristic of the methods of proving independence and the consistence of axioms, and from the examples analyzed above, that in this procedure we accept as true or false certain sentences to which we reduce given inscriptions by corresponding interpretation of undefined constants.
>
> For instance, we have accepted as true the sentence: "if X is a point and Y is a point, then some straight line contains X and Y" ...
>
> To round off information on formalism in the building of a deductive system, we may add that such formalism recommends itself by a certain economy of work. By applying a formal procedure in proving theorems of a given discipline we may thereby also prove theorems of some other discipline, if the objects investigated by the latter so correspond to the objects investigated by the former that sentences of the same form, according to the interpretation of the constants they include, state theorems either of the former or of the latter discipline [Kotarbiński, 1966, 256].

This contrasts with Hilbert in a way broadly in tune with Padoa: unlike Hilbert, the primitive terms of an axiom system are associated with ideas not determined by the axioms, and it is simply an intriguing sort of generality that it can turn out that the same axiom system can receive different intuitive interpretations. The system bears an intended interpretation not settled by its structure, though generality can be achieved by varying the interpretation imposed upon it.

Tarski's own remarks on formalism have a similar flair. Consider first this one, from Sur La Methode Déductive [Tarski, 1986f], which was later incorporated into Tarski's logic textbook:[23]

> As is known, every mathematical science is founded upon a suitably chosen system of primitive terms and axioms. Our knowledge of the objects designated by the primitive terms could be quite extensive, and the accepted axioms need not exhaust it. Yet this knowledge is our private affair, exercising not the least influence upon the construction of the science. Thus, in deriving a theorem from the axioms, we make no use of this knowledge—we conduct ourselves as though we did not understand the sense of the concepts with which we occupy ourselves in these considerations, as though we knew nothing beyond

that which was explicitly stated in the axioms; we abstract from the sense of the primitive terms we admit and we attend to nothing but the form of the axioms in which these terms appear [Tarski, 1986f, 331] (my translation).

Interestingly, in the English version of the passage in the textbook of 1941, Tarski hedges by saying that our grasp on the primitive terms is "so to speak" our private affair, while in the German version the declaration is unhedged, as in the French, but the German asserts that the axioms "must not" exhaust our knowledge (*muß keinesfalls durch die angenommenen Axiome erschöpft werden*).

Padoa anticipates some aspects of Intuitionistic Formalism. The insistence that the intended interpretation of a theory is somehow important, but also strictly irrelevant to the theory itself is one important point here. We have already seen exactly this emphasis on the separability of a deductive theory from its intended interpretation in the opening pages of ODS [Tarski, 1983f, 112]. Perhaps even more important is Padoa's entirely psychologistic take on "interpretation", which here as elsewhere in the article is clearly something "in the mind".[24] This is something to which Leśniewski and Kotarbiński were entirely sympathetic, and Tarski's talk of our knowledge of the denotations of the primitive terms being "our private affair" shows that he was, too.[25] To return again to our contrast from the introduction, Intuitionistic Formalism is a view within expressive rather than representational semantics.[26]

It is sometimes argued (e.g. [Hodges, 2009, 482]) that Intuitionistic Formalism is actually incompatible with recognizing that an axiom system can be satisfied by many models. This isn't the case. In fact the doctrine took this fully into account. Recall the psychologistic aspect of Intuitionistic Formalism that is already at work in Padoa and continues into Leśniewski. Notice that Leśniewski talks of "placing" or "putting" determinate senses on the theses of his system, and of these theses, taken with these meanings, having "intuitive validity" *for him*. The picture is one of Leśniewski having put himself (and perhaps others) in a position to associate certain thoughts with certain formulae of the system, the system in turn being designed in such a way that inference according to its rules always leads from sentences that express intuitively valid thoughts to others that do so as well via intuitively valid inferences. There need be no idea here that the system couldn't bear other interpretations as long as the inferential relations among the other thoughts were the same. We will see more about this psychologism and how it affects the doctrine, especially when we look at Tarski's understanding of Padoa below (§ 2.2.3).

1.4.2 Monism vs tolerance

Leśniewski, intended the thoughts expressed in his formal systems to capture the true nature of things; see [Leśniewski, 1992h, 177], discussed above, on "non-intuitive", merely non-contradictory systems as contrasted with an "intuitive" mathematics that is "directed resolutely towards reality", as well as his various disparaging remarks about formalism. Leśniewski's systems, and the theory of semantical categories on which they rested (see § 4.2.2), were supposed to be *the* correct system of logic and foundations of mathematics. By the time of *Logical Syntax* there thus comes to be a major difference between Carnap and Leśniewski on what Carnap calls the "principle of tolerance". In 1929 the Carnap of the *Abriss* works entirely within the STT:[27] "Logistic" just is simplified type theory directed toward warding off paradox by prohibition of "level-mixing" [Carnap, 1929, 20], but it is also a "sharp tool" for the analysis of assertions and concepts in every domain [Carnap, 1929, iii]. By 1934 Carnap had moved to a forthrightly pluralistic attitude toward logic:

> The following explanations are concerned with the symbol '=' considered as the symbol of identity in the narrower sense (that is to say, as used between ℨ or between object-designations) and not the symbol of equivalence (that is to say, as used between ℭ). The symbol of identity occurs in Languages I and II (and also in the languages of Frege, Behmann, Hilbert) as an undefined symbol. Following Leibniz, Russell defines '$x = y$' in the following way: "x and y agree in all their (elementary) properties." Wittgenstein rejects the symbol altogether and suggests a new method for the use of variables by which it may be avoided.
>
> Philosophical discussions concerning the justification of these various methods seem to us to be wrong. The whole thing is only a question concerning the establishment of a convention whose technical efficiency can be discussed. No fundamental reasons exist why the second or third of these methods should not be used instead of the first in Languages I and II [Carnap, 2002, 49].

> Our attitude toward requirements of this kind is given a general formulation in the *Principle of Tolerance: It is not our business to set up prohibitions, but to arrive at conventions* [Carnap, 2002, 51].

> *In logic, there are no morals.* Everyone is at liberty to build up his own logic, i.e. his own form of language, as he wishes. All that is required of him is that, if he wishes to discuss it, he must state his methods clearly, and give syntactical rules instead of philosophical arguments [Carnap, 2002, 52].

As Ricketts notes, with this Carnap loses the connection to the expression of thoughts that was so central to Frege's project in Logic:

> In repudiating the ideal of a unique, science-embracing logic, Carnap rejects Frege's assumption of a common store of logically interrelated thoughts expressed by the sentences of colloquial language and perspicuously expressible by sentences couched in the framework of *Begriffsschrift*. Carnap cannot then rely on any overarching notion of content to give his syntactic investigations their application to actual or hypothetical languages for science [Ricketts, 2004, 191].

> We can now appreciate the depth of Carnap's rejection of Frege's conception of a thought as that for which the question of truth arises. Linguistic behavior is, so to speak, in itself logically amorphous. We bring logic to it by coordinating a calculus with it. The scientific philosopher, the logician of science, describes various calculi. She then can freely pick any of these calculi, and envision her group to speak a language coordinated with it. In this way she applies to her group's hypothetical utterances, the syntactic explications of epistemic notions that the syntactic description of the calculus makes available. Application of terms like "true," "false," "consequence," and "consistent" in the logic of science become tolerably precise only via such coordination ... This is how tolerance in logic is possible [Ricketts, 2004, 193].

Leśniewski, and with him Kotarbiński, would not approve: for them the whole point of designed conventions of language was to faciltate the clear expression of thoughts that were independent of the system and assumed to be clearly grasped. Recall in this context that Frege is the only predecessor for whom Leśniewski expresses much respect in [Leśniewski, 1992h]. We will see Tarski's view change on this issue between the body of CTFL and the Postscript. What is of significance for us at this juncture is simply the strongly epistemic conception of logical consequence that is at the heart of Intuitionistic Formalism as Leśniewski conceives of it. A deductive theory, for Leśniewski, is a tool for the expression of thought and reasoning. It isn't an object of study in its own right, and it must thereby be crafted to respect the logical relationships among thoughts intuitively judged.

1.4.3   Five doctrines

We have now seen five main aspects of Intuitionistic Formalism in Leśniewski and in the related philosophy of language of Kotarbiński:

1. The understanding of language as a medium for the expression of thought via a speaker's intentions.
2. The conception of conventions of language as determining which thoughts are expressed by the use of a given sentence when a language is used correctly.
3. The conception of axioms and theorems as sentences that are assertible in the sense of seeming true, given the conventions, to those who are party to them.
4. The conception of the intuitive meaning or content of a term in terms of the traditional logical notion of connotation.
5. The goal, in theory construction, of constraining the interpretations of the primitives as much as possible through the implication of theorems that constrain the assignment of intuitive meaning to the terms of a theory.

All of these aspects of the view remain in Tarski's work of the period in which we are interested, save that in his conception of the fourth he makes no reference to doctrines about connotation from traditional logic.[28] He keeps, however, the idea that something like conceptual analysis traditionally conceived is possible and important, and his definitions are intended to answer, though not in the way usually assumed, to such analyses. Let us here briefly marshal some *prima facie* evidence that these views are to be found in Tarski's works of the period.

(1) *Language as Expressive* That language is a medium of expression via speakers' intentions is, as we saw, an important doctrine in Twardowski which is worked out more thoroughly in Kotarbiński's account of direct statement. Now Tarski writes that "normally expressions are regarded as the product of human activity" and links this to the issue of nominalism [Tarski, 1983a, 174, note 2]. The "human activity" here can only be expressing thoughts as understood by Kotarbiński. The reference to actions and products, as well as to nominalism, are clear references to Twardowski and Kotarbiński and Tarski's sheepishness about nevertheless setting the point aside for the sake of doing logic indicates that he still accepts these views as essentially right.

Another indication of Tarski's sharing this orientation comes from a point obliterated by Woodger's translation. Consider this passage:

> Now formalized languages have hitherto been constructed exclusively for the purpose of studying *deductive sciences* [Tarski, 1983a, 166].

As Hodges [Hodges, 2008, 98] notes, Woodger's "studying deductive sciences" renders Tarski's "*uprawiać sformalizowane nauki dedukcyjne*" [Tarski, 1933, 16], and "*uprawiać*" means not *study* but rather *practice*. Blaustein has this right with "*formalisierte deduktive Wissenschaften zu betreiben*" [Tarski, 1986a, 72]. The idea is essentially Leśniewski's, as Hodges discusses:

> Tarski refers several times to 'practising' or 'performing' a deductive theory; sometimes he adds 'on the basis of its language'... There is really only one activity he can have in mind. This is the activity of sitting in front of a piece of paper and writing a sequence of formulas in the language of the theory, where each formula is either an axiom of the theory, or a definition, or something derived from previous rows by rules of the theory.
>
> There are not many published examples of this activity... For virtually pure examples one should look at papers of Leśniewski... [whose paper] on Abelian groups is eleven pages long, and nearly eight pages consist of a formal derivation using nothing outside of the formal language and symbols to index the applications of rules [Hodges, 2008, 97].

> The working mathematician proves things informally at first. But proofs need to be checked and communicated to other people. For these purposes mathematicians have agreed on certain ways of formalizing their intuitive thoughts, namely as deductive theories. They have agreed, for example, that we should take as primitive terms of our deductive theories expressions that 'seem to us to be immediately understandable', and we should take as axioms 'statements whose truth appears to us evident' ... The ideal mathematician sets up deductive theories that meet these requirements and then 'practises' these deductive theories.
>
> I believe that the previous paragraph summarizes what Leśniewski meant when he called himself an Intuitionistic Formalist, and what Tarski meant when he said in 1930... that this was his position, too. [Hodges, 2008, 103].

This further explains why the expressions of the deductive theories we study in defining truth have to be meaningful: the theories are actually to be used to express intuitively true thoughts. What is interesting here is that although Tarski, as a metamathematician, *is* interested in *studying* deductive sciences, as a philosopher his view is that the point of a deductive science is that it is to be used clearly to express thoughts, just as Leśniewski maintained. Indeed, as we will see in the remainder, in the late 1920s through the early 1930s Tarski's aim in studying formalized languages was to study the conditions under which the goals set for their use so conceived were met. It is to this end that he conceived the project of seeing to what extent the metamathematical notions needed to study and evaluate a deductive science could themselves be incorporated within a deductive science that could be used rigorously to express metatheoretic thought. Thus, since Tarski holds that deductive theories are to be "practised" he has to be thinking of them in the first instance as vehicles for the expression of thought.

Commentators often note familiar contrasts between logic conceived of as a medium for the expression of thought, and logic, or logical systems, as objects of study and apply them to Tarski. Sundholm [Sundholm, 2003] stresses the distinction between what he calls "languages with meaning" and "languages without use", where the former are languages designed, in the tradition of Frege, to be actually used to express mathematics and the latter are, in the tradition of Thomae and Hilbert, to be objects of metamathematical study. Sundholm characterizes Tarski's progression during the period from the formulation of CTFL in 1929 to the Postscript of 1935 as a shift from a Leśniewskian adherence to the first tradition to a more mathematical attachment to the second. One thing I will do here is to stress that, to the extent that the two perspectives adequately characterize the endpoints of Tarski's development during the period, there was an intermediate phase, and all of the important work occurs during it: namely, a phase in which Tarski tried to blend both traditions by designing deductive theories situated within the first tradition precisely to be used for the second sort of activity. The same point thus applies, *mutatis mutandis* to the related characterizations of two perspectives on language of [Hintikka, 1996] (*lingua universalis* vs *calculus ratiocinator*) and [van Heijenoort, 1967] (logic as language vs logic as calculus). I also date the abandonment of the earlier perspective, in the form on Intuitionistic Formalism, later than Sundholm does, since I find it not in the 1935 Postscript to CTFL but in the definition of consequence in terms of models, since the postscript is still committed to a derivational conception of consequence, central to Intuitionistic Formalism, which is overturned only in "On the Concept of Following Logically" (CFL).

(2) *Convention* One significant piece of evidence here is the oft-noted fact that Tarski calls Convention T a "Convention" (*Konvention, umowa*). This accords perfectly well with Leśniewski's view that a definition is a kind of convention for the use of a term. Now the appeal to convention can be expected to be somewhat indirect, since in Leśniewski's later work the appeal to convention is taken up in formation rules and directives of his systems: rules for the formation of sentences, and rules for the assertion of one sentence on the basis of others. In light of the earlier work, however, we can see that such rules are conceived of as playing the role of laying down principles for the use of signs that suit them to express various intuitive contents.

Throughout the articles with which we are concerned Tarski of course thinks of deductive sciences as precisely governed by such rules [Tarski, 1983f, 113–116], [Tarski, 1983a, 166]. Since Tarski is concerned with the expression of intuitive meaning in deductive sciences as Leśniewski was, his conception of such rules would have been along Leśniewski's lines as well, and since Leśniewski's conception of language and meaning was allied with Kotarbiński's, Tarski's conception of the relation between the rules governing sentencehood and theoremhood in a system would have been in accord with Kotarbiński's account of the role of convention in determining what is "indirectly stated" by a sentence.

One further piece of apparent evidence requires more delicate handling. Tarski uses certain phrases that come out in Woodger's translation sounding a good deal like the English translation of Kotarbiński's *Elementy*. For example, in CTFL we find this familiar passage:

> We take the scheme (2) and replace the symbol 'x' in it by the name of the given sentence, and 'p' by its translation into the metalanguage. All sentences obtained in this way, e.g. '$\bigcap_1 \bigcap_2 (\iota_{1,2} + \iota_{2,1})$ *is a true sentence if and only if for any classes a and b we have $a \subseteq b$ or $b \subseteq a$*', naturally belong to the metalanguage and explain in a precise way, in accordance with linguistic usage, the meaning of phrases of the form 'x is a true sentence' which occur in them [Tarski, 1983a, 187].

This seems to compare favorably to passages like this in Kotarbiński:

> A given phrase states a certain experience indirectly whenever, without stating it directly, such can, in view of linguistic usage or adopted conventions (*na obyczaj tego jkęzyka lub ustalone co do niego umowy*), be used in a given language as a direct statement of such an experience [Kotarbiński, 1966, 4] [Kotarbiński, 1961, 14].

Unfortunately this is one of the points at which Tarski is actually working behind the scenes to bury the pervasive appeal to intuition found
in the original Polish of CTFL. The Polish phrase that Woodger has as
"in accordance with linguistic usage", following Blaustein's "*mit dem
Sprachgebrauch übereinstimmender*" [Tarski, 1986a, 45], is "*zgodny z intuicjką*" [Tarski, 1933, 39]: consistent with intuition. (The same switch can
be found on the previous page.) This is of course helpful to us in arguing that Tarski was concerned with intuitive meaning in the original
work, but it renders problematic the current attempt to gather evidence
that Tarski conceived of linguistic expressions as significant because they
are governed by conventional rules that determine what intuitive contents they express. Since Blaustein doctored the translation under Tarski's
direction I think we can be fairly confident that the German wording
that Woodger follows is Tarski's. But since Tarski was happy with cognates of "intuitive" in French and Polish, but not in German and English
(cf. § 4.1.1), the motivation for the change in the German translation
was presumably the desire not to raise hackles among German readers.
(For another example of Tarski's strategic "translations" from Polish to
German, see § 4.3.3.)

So the question is whether "*mit dem Sprachgebrauch übereinstimmender*"
was conceived of by Tarski as appealing to the conventionalist ideas
expressed in Kotarbiński's "*na obyczaj tego jkęzyka lub ustalone co do niego
umowy*", or was rather a phrase he knew from some German-language
author that he thought would go down better in Germany and Austria
than "*mit den Intuition übereinstimminder*". Certainly the agreement with
Kotarbiński's view would have been manifest to Tarski and the appeal to
linguistic usage would also likely have struck him as a good substitute for
the appeal to intuition, backed up by Kotarbiński's views, since in well-
designed conventions linguistic usage *does* correspond to and express our
intuitions about what expressions mean. So I am inclined to think that
this is some evidence that Tarski accepted Kotarbiński's notion of convention, but matters in that respect aren't entirely clear. What is clearer
is that the wording is Tarski's, since Blaustein isn't otherwise so free with
his translations, so we can grant that Tarski was happy expressing himself
by appeal to the idea that what he was trying to capture in a deductive
science was the ordinary usage of phrases, and surely here incorrect usage
wouldn't have been to the point, so there is some normative notion close
to the appeal to convention packed into the phrase. In any case, in the
end the appeal to convention here isn't terribly controversial as ascribed
to Tarski: if one thinks of an axiomatic theory as intended to have theorems that convey certain thoughts, one has to think of it as governed

by stipulations as to how expressions are to be used, and these are just rules for the use of the sentences upon which users of the system agree.

(3) *Assertability* It is crucial to keep in mind that Tarski takes the meaning of the primitives of a deductive science to be settled by which sentences are *treated as* true, not which sentences actually are true [Ray, 2003]. Tarski's conception of the meanings of primitive terms is part of a general theory of deductive theories—sets of sentences closed under some assumed relation of consequence [Tarski, 1983d, 40]—the generality of which would be pointlessly compromised if the sentences in question had to be true. Whenever the claim that primitive terms have their meanings settled by axioms is under discussion, Tarski talks about sentences as being "held true" or of what "can be asserted", e.g. [Tarski, 1983a, 167], [Tarski, 1983k, 299] and [Tarski, 1983k, 299, note 1].[29] The claim that primitive terms have their meanings established by sentences that are *held* true rather than by sentences that *are* true is crucial to understanding Tarski's conception of an inconsistent language, as we will see later in § 5.2.2.

(4) *Content and Connotation* Tarski speaks often of defining "the content" of a concept, rather than just defining a concept [Tarski, 1983a, 187], [Tarski, 1983b, 401] [Tarski, 1983h, 409]. At [Tarski, 1983f, 112] we have the "content" of the term "definable set of real numbers", but we know from Kotarbiński that this is also the content of the concept of a definable set of real numbers. This locution can look to the contemporary reader like a verbal hiccup, but it is in fact an acknowledgement of doctrines about the content of mental acts hailing from Twardowski and accepted by Kotarbiński and the early Leśniewski. On this view, a "concept" is a mental particular, while its "content" is what in the period might have been called its "intentional object". Tarski's "contents of concepts" are therefore fairly close to what the contemporary reader would mean by "concept" itself. Tarski appears to have found the locution uncomfortable, because he also sometimes treats "content of the concept" and "the concept" as synonymous—a definite no-no by Twardowski's standards, but one that matches better the use of "concept" outside of Poland. Nevertheless, the appearance of the locution, and the treatment of the content of a concept as what a definition is supposed to capture, indicate that the basic ideas at work are the ones we have found in Leśniewski and Kotarbiński.

(5) *Capturing Contents* Passages on the way in which axioms and theorems express properties of concepts or determine meanings can be found at [Tarski, 1983g, 31], [Tarski, 1983c, 63] and [Tarski, 1983f, 112]. Our main task in the next two chapters will be to establish that Tarski's project

was the capture of the contents of metamathematical concepts within deductive metatheories, but the cited passages make at least the initial case that Tarski is thinking of things along Leśniewski's lines.

Having in hand a good idea of what Intuitionistic Formalism really was for Tarski, we can now read a familiar passage in an unfamiliar but much more illuminating way:

> It remains perhaps to add that we are not interested here in 'formal' languages and sciences in one special sense of the word 'formal', namely sciences of the signs and expressions to which no meaning (*inhaltlich Sinn*) is attached. For such sciences the problem here discussed has no relevance, it is not even meaningful. We shall always ascribe quite concrete and, for us, intelligible meanings (*Bedeutungen*) to the signs which occur in the languages we shall consider [Tarski, 1983a, 166–7].

It is generally thought about this passage that it expresses commitment to the idea that the concept of truth will be studied in languages that are "interpreted" in the model-theoretic sense. We can now see, however, that the passage is actually an expression of Intuitionistic Formalism. The "concrete, intelligible meanings" at issue aren't model-theoretic interpretants, but intuitive concepts.[30] Substantiating this claim will have to wait upon a fuller understanding of the manner in which, in 1931, Tarski simply couldn't have been thinking of meaning in terms of his semantics.[31]

1.4.4 Tarski's project

As noted above, when it comes to the relation between the intuitive concept intended to be expressed by a term and the formal role instituted for that term by the structure of an axiomatic theory, the possibility of various kinds of mismatch and failure threatens: the conventions governing the use of a sentence involving a term intended to have a certain connotation might fail to constrain the association of that connotation with the term down to the intended one. A project thus beckons: investigating the conditions under which the usage of terms is adequate to the connotations grasped in the concepts with which they are associated.

A simple example will illustrate the point; we will meet Tarski's own applications soon enough. Consider three predicates, *F*, *G* and *H*, intended to connote the properties of being scarlet, being red, and being blue, respectively, and suppose for the sake of example that it is some sort of conceptual truth that anything scarlet is red. If, according to the conventions of a language, $(\forall x)(Fx \rightarrow Gx)$ isn't assertable then the connotation of F hasn't been fully captured, since anyone with the concept of being scarlet will accept that anything scarlet is red, while if $(\exists x)(Fx \,\&\, Hx)$ is assertable, then the connotation of F hasn't been captured at all, since anyone who possesses the concept of being scarlet will accept that nothing scarlet is blue. Only a language in which $(\forall x)(Fx \rightarrow Gx)$ is assertable but $(\exists x)(Fx \,\&\, Hx)$ is not is a language that does the Leśniewskian job of representing the intuitive connections among thoughts expressed when F, G and H are intended to express the concepts of being scarlet, being red, and being blue.

So we have two questions about any given term in a deductive theory: are the conventions governing the assertable sentences in which it appears sufficient to express its intended connotation, and are the conventions governing them consistent with its expressing its intended connotation? The requirement that these questions be answered in the affirmative is, in embryo, the "material adequacy" requirement of Convention T. [32]

The project, then, is to craft a deductive theory that forces, though its theorems, its primitive terms to express the contents of certain concepts. This is what remains in Tarski's work of the appeal to connotation in Leśniewski and Kotarbiński. The usual sort of conceptual analysis is what Tarski refers to when he remarks that the reader is familiar with the concept of truth and can find it discussed in works on the "theory of knowledge" at [Tarski, 1983a, 153] and it is what he refers to when he says that he is concerned with the "content" of "definable set of real numbers" and will remind the reader of it at [Tarski, 1983f, 112]. Then the point that arises from Leśniewski's treatment is that the role of a sign intended to express a concept within a deductive theory, determined by conventions, must capture its intuitive content.

In what way is the content of a concept to be captured in a deductive theory? The simplest way is for a statement of the analysis to be among the theorems of the theory. This, as we saw (§ 1.2.2) was Leśniewski's way of settling the interpretation of "ε" in his Ontology [Leśniewski, 1992h, 367–8]. We will find an example of this in the next chapter in Tarski's early treatment of the consequence relation. But the relationship need not involve an explicit statement of the analysis appearing among the theorems: what is required is simply that accepting the conventions and accepting the axioms as true constrains a rational thinker to associate the right content with the terms of the theory. Later we will see that Tarski's treatment of the content of the concept of truth exemplifies this inexplicit form of capture: the T-sentences being theorems of the metatheory, given the interpretations of the other expressions of the object- and metalanguages will force "is true" to express truth conceived according to the "semantical definition" [Tarski, 1983a, 155] even though the T-sentences aren't themselves part of or implied by this analysis of truth.

An ideal deductive theory, from an Intuitionistic Formalist perspective, would be one maximally immune from misunderstanding, a theory such that two thinkers who contemplated it, accepted the conventions governing the conditions under which sentences are assertable, and accepted its theorems as true would have to associate the same concepts with its terms. In most cases, for obvious reasons, this is unattainable: the theory has primitive terms, and the interpretations of these can be varied; indeed, we have seen that it can be of some value, in Kotarbiński's phrase, for an "economy of work". But at least the following is a goal: the axioms and primitives should be kept to a minimum so that the work of securing agreement among interlocutors on intuitive meaning is made as simple as possible, and as much as possible that is of interest in the domain in question should be settled completely once the axioms are accepted. The limiting, ideal case, on Tarski's view, however, is a theory that does uniquely determine the concepts expressed by its primitive terms (§ 2.3).

Tarski, having accepted Intuitionistic Formalism, set out to study the extent to and the conditions under which this could be achieved. But Tarski also focused on a particular application: "metamathematics", conceived of as the discipline in which deductive theories whose intuitive subject matter was other deductive theories are used:

> The deductive disciplines constitute the subject-matter of the *methodology of the deductive sciences*, which today, following Hilbert, is usually called *metamathematics*, in much the same sense in which spatial entities constitute the subject-matter of geometry and animals that of zoology ...
>
> The present studies ... aim is *to make precise the meaning of a series of important metamathematical concepts* which are common to the special metadisciplines, *and to establish the fundamental properties of these concepts* [Tarski, 1983c, 60]

Tarski wants to "make precise the meaning" of concepts which are deployed in metatheoretic thought, and to do so within deductive theories. The need to bring metamathematics up to the standards of Leśniewskian Intuitionistic Formalism meant that Tarski had to begin to formalize things that previously went unformalized, and he was proud of this achievement: "as far as I know the metatheory has never been given in the form of an axiomatized system" [Tarski, 1983a, 173 note 3].

In the "Historical Notes" Tarski claims independent discovery of "the axiomatic construction of the metasystem... and in connexion with this... the discussions on pp. 184 f. on the interpretation of the metasystem in arithmetic... which was developed far more completely and independently by Gödel" [Tarski, 1983a, 277–8]. We turn next to Tarski's first contribution to this project.

2
Tarski as Intuitionistic Formalist

Accepting the basics of Intuitionistic Formalism, Tarski sought to extend it. In particular, inspired perhaps by Leśniewski's steps toward formalized syntax [Leśniewski, 1929], [Leśniewski, 1992e], Tarski began to look at the issues that arose when one's discourse *about* a deductive theory was itself expressed in a deductive theory. This is the sort of project that would have attracted an ambitious young contributor: investigate whether the going view among his teachers about "metamathematics" could handle metamathematical concepts themselves. His first published contributions to this project were several papers on axioms for the consequence relation. We will look at those works in order to see the beginning of Tarski's project. After that we will look at his most extensive contribution to the general project, "Some Methodological Investigations on the Definability of Concepts", which collects work from 1926 and 1932.

2.1 The early metamathematical works

2.1.1 Axiomatizing consequence

The aim of the papers collected as essays III to V of *Logic, Semantics, Metamathematics* is to capture axiomatically the central concept of logic: consequence, one sentence's following from some others.[1] The papers will be of interest to us both because they provide an early example of the Intuitionistic Formalist project and because the treatment of the consequence relation is one of the most important aspects of Tarski's development in the 1920s and 1930s. Tarski develops extensional semantics as a contribution to the Intuitionistic Formalist analysis of metamathematical concepts, but well into this project, as we will see— as late as 1934–5, in fact—Tarski remains wedded to a conception of

53

consequence on which for one sentence to follow from others is simply for it to be derived from them through intuitively valid rules of inference. Tarski states his goals in the first of the three papers as follows:

> Our object in this communication is to define the meaning (*den Sinn ... zu präzisieren*), and to establish the elementary properties (*elementaren Eigenschaften festzustellen*), of some important concepts belonging to the *methodology of deductive sciences*, which, following Hilbert, it is customary to call *metamathematics ...*
>
> Formalized deductive disciplines form the field of research in metamathematics roughly in the same sense in which spatial entities form the field of research in geometry. These disciplines are regarded, from the standpoint of metamathematics, as sets of *sentences*. Those sentences which (following a suggestion of S. Leśniewski) are also called *meaningful sentences* (*sinnvolle Aussagen*), are themselves regarded as certain inscriptions of a well-defined form ... From the sentences of any set X certain other sentences can be obtained by means of certain operations called *rules of inference*. These sentences are called *consequences of the set X*. The set of all consequences is denoted by the symbol '$Cn(X)$'.
>
> An exact definition of the two concepts, of sentence and of consequence, can be given only in those branches of metamathematics in which the field of investigation is a concrete formalized discipline. On account of the generality of the present considerations, however, these concepts will here be regarded as primitive and will be characterized (*charakteriziert*) by means of a series of axioms. In the customary notation of general set theory these axioms can be formulated in the following way:
>
> Axiom 1. $\bar{\bar{S}} \leqslant \aleph_0$.
> Axiom 2. *If* $X \subseteq S$, *then* $X \subseteq Cn(X) \subseteq S$.
> Axiom 3. *If* $X \subseteq S$, *then* $Cn(Cn(X)) = Cn(X)$.
> Axiom 4. *If* $X \subseteq S$, *then* $Cn(X) = \sum_{Y \subseteq X \text{ and } \bar{\bar{Y}} \leqslant \aleph_0} Cn(Y)$.
> Axiom 5. *There exists a sentence* $x \in S$ *such that* $Cn(\{x\}) = S$. [Tarski, 1983g, 30–31] [Tarski, 1986i, 311]

The passage conveys the basic Intuitionistic Formalist ideas. The appeal to Leśniewski in distinguishing a closed well-formed formula as a "meaningful" sentence is one obvious point. Next, there is a concept of consequence that can be captured within a variety of deductive theories but which is independent of them—it can be defined within one theory and axiomatized within another. Note, relatedly, that "defining" thereby has to mean something other than "analyzing" in a sense we would recognize (cf. § 1.1.1), since a particular definition of consequence within the metatheory for a single deductive theory hardly counts as a general analysis of the concept of consequence. Furthermore, the basic way a concept is "characterized" within a deductive theory is to have the term that expresses it constrained by a set of sentences held true. In laying down his axioms, Tarski is laying down conventions for the use of his terms that determine that "$Cn(X)$" has to express the content of the concept of consequence.

The exercise is, then, simultaneously philosophical and formal. The philosophical aspect of it, though reduced to a minimum in the case at hand, is articulating and defending an analysis of the concept of consequence. Tarski takes the analysis of the concept of consequence to involve the idea that consequence is transitive, that a set of sentences is among its own consequences, and so forth. These are not controversial claims when it comes to consequence and so Tarski does not tarry over them, but the basic intuitionisitic formalist structure is in place in these articles: we express a concept within a deductive theory by laying down axioms that are sufficient to capture its associated conception, in this case directly by implying a statement of it.[2]

One important point about the analysis of consequence here needs to be taken note of. It could appear from Tarski's idea that there is a general notion of consequence which can only be worked out in detail in a particular deductive system that his conception of consequence is, at the time of writing before 1928, actually independent of the rules of proof and hence somehow non-epistemic: he has, it might appear, a conception of one thing following from others that isn't tied to particular "transformation rules". This appearance is, however, misleading. What Tarski assumes in this passage is that we have is a general conception of one thing's following from others by rules that we leave unspecified. It is still the case here that to follow from is to follow by rules of proof. This is important for us to see, since when we begin to look at Tarski's development of extensional semantics we will have to note that he does not see the possibility of understanding consequence itself in terms of it until very late and, as we will discuss in § 7.3.2, though people were studying models of theories in the 1920s and earlier, nobody, including Hilbert and Gödel, was thinking of logical consequence itself other than derivationally.

2.1.2 Relativization to a deductive science

We now come to a significant question about what happens to Intuitionistic Formalism in Tarski's hands. As Leśniewski understood it, the goal in the construction of a deductive theory was to craft an axiomatic structure that constrained the contents of the concepts expressed by the primitive terms of the theory as tightly as possible—ideally down to uniqueness, though that goal is not achievable for most concepts and theories. Protothetic, Ontology, and Mereology are supposed to be so tightly constructed that they zero in on particular concepts; Ontology, for instance, is supposed to come as close as possible to singling out *the* one and only concept of one thing's bearing the ε-relation to another.

Tarski, by contrast, is concerned with concepts that have an essential relativization to a language. A particular deductive theory concerning the consequence relation, whether it be axiomatic or whether it introduce "$Cn(X)$" by definition, is restricted to a particular object language:

> Strictly speaking metamathematics is not to be regarded as a single theory. For the purpose of investigating each deductive discipline a special metadiscipline should be constructed. The present studies, however, are of a more general character: their aim is *to make precise the meaning of a series of important metamathematical concepts* which are common to the special metadisciplines, *and to establish the fundamental properties of these concepts*. One result of this approach is that some concepts which can be defined on the basis of special metadisciplines will here be regarded as primitive concepts and characterized by a series of axioms [Tarski, 1983c, 60].

In addition to stating the basic Intuitionistic Formalist ideas, this passage invites a question that is very familiar from the literature on Tarski's truth-definitions: how can Tarski claim to define "the" concept of consequence in a metatheory that concerns only a single object language? The related question about the concept of truth and his particular truth-definitions has been a source of consternation for decades: how can Tarski claim to define "the" concept of truth in a theory that concerns only a single object language? We should be given pause here, however, by the fact that even Tarski's treatment of the lowly concept of a sentence raises exactly the same questions. In the early articles on the consequence relation he simply takes the notion of a sentence (closed formula) for granted, but when he defines being a sentential function and then being a sentence at Definition 10 and 12 of CTFL [Tarski, 1983a, 177–8] his definition raises every question raised by the much more thoroughly discussed Definition 23.[3]

One way to put the obvious problem is this. Suppose we grant that "S" is given some axiomatization or definition in the metatheory that suits it to apply to all and only the sentences of the object language L. Why hold on this basis that "*S*" expresses "the" concept of a sentence? After all, for all that the theory settles, it could either express the concept of being a sentence in general, or the concept of either being a sentence of *L* or a seven character string of some other language (or being either a sentence of *L* or a turnip, or...). How can Tarski hold that constraining a term by some theorems in a theory that applies to only one language settles the issue that one concept is thereby expressed as opposed to others that agree with it as applied to that language?

Metamathematics, as Tarski construes it, is concerned with "formalized deductive disciplines" [Tarski, 1983g, 30] [Tarski, 1983c, 60] or, what comes to the same thing for him, "formalized languages" [Tarski, 1983a, 165]. "These", he writes, "can be roughly characterized as artificially constructed languages in which the sense of every expression is uniquely determined by its form" [Tarski, 1983a, 165–6]. The lack of this central feature is likewise what makes unformalized disciplines or languages unsuited for study [Tarski, 1983c, 60]. An assumption of metamathematics, then, is that expressions differ in form if they differ in meaning. Nevertheless Tarski follows Leśniewski in taking the expressions of the language he studies to have intuitive meaning [Tarski, 1983c, 62]: "the expressions which we call sentences still remain sentences after the signs which occur in them have been translated into colloquial language" [Tarski, 1983a, 167]. The point is easy to miss, but at the outset of [Tarski, 1983g] and [Tarski, 1983c], and not far into [Tarski, 1983a] Tarski has already established to his satisfaction that although meaningful language is the object of study, expressions and sentences will be treated as meaningful only insofar as their syntactic forms can go proxy for their meanings.

Now one notion for which the form of an expression *cannot* go proxy is the notion of an expression's being in a language [Tarski, 1983a, 264]. Though expressions get their meanings from the conventions of language—in the case of formalized languages, formation rules and rules of proof, rules of definition and stipulations of axiomatic status [Tarski, 1983c, 63]. [Tarski, 1983f, 112–13] and especially [Tarski, 1983a, 166]— the notion of an expression's being in a language simply isn't suitable for metamathematical treatment, despite Tarski's acknowledgement of the dependence of meaning on rules of language in passages such as [Tarski, 1983a, 153]. Hence Tarski always depends on the reader to keep in mind that sentencehood is always understood relative to the particular object language for which a given metatheory is set up, with its conventions assumed to be in place. In this respect, the relativization to the object language, when it comes to associating an intuitive meaning with the relevant terms of the metalanguage, is left in the mind of the reader in the same way as is the interpretation of the primitives of the metatheory to the extent that the theory doesn't settle it. Here as elsewhere (e.g., as with Padoa's method, as we will see in § 2.2.3) Tarski's goal is to get as much as possible explicitly on the page, but not everything can suitably be put there. (This, by the way, explains the puzzling fact that Tarski requires that each metatheory concern only a single object language [Tarski, 1983c, 60]: if a metatheory was about more than one object language, it would have explicitly to involve the non-formalizable notion of being in a language.)

We will do well, then, to settle these matters now, focusing on the simple concept of a sentence (or, in current terms, a closed well-formed formula) and then turning to consequence. The reason that Tarski doesn't try to give an absolutely general definition of a sentence in terms of the form of expressions is perfectly clear: to be a closed well-formed formula is just to be declared such by the the formation rules of a given language. Something is a sentence only relative to a language, and since Tarski, following Kotarbiński and Leśniewski, thinks of a language in terms of conventions, only relative to a set of conventions. Equally clear is why Tarski doesn't give a definition of a sentence of L for variable L: here there is again nothing to the definition other than that to be a sentence is to be declared one by the formation rules of some language or other.

Less clear, however, is why Tarski suppresses reference to the relativization of sentencehood to a particular language, and even less clear than that is why he is willing to talk about axiomatizing or defining one and the same concept of a sentence relative to any particular language or, generally, to some language or other, as he clearly is in the above passage. Since the extension of "sentence" differs by language, one might well think that the concept it expresses does, too. The parallel with the discussion of truth here is clear: Tarski doesn't define truth in L for variable L, and when he does define truth relative to a particular object language he suppresses this relativization, yet glibly retains talk of "the" content of the concept of truth.

What we have here, then, is a question about relativized concepts. To take a reasonably apt comparison, being an illegal act is presumably relative: there is no such thing as doing something illegal *simpliciter*, there is only doing something illegal relative to some set of laws or other. We thus have two options: we can hold that there is a single concept of an act's being illegal and that the relativization to a set of laws required to determine an extension for it is "external" to the concept, or we can hold that there really isn't a general concept of being illegal and that there is, rather, only a set of concepts of being illegal according to the laws of New York State, being illegal relative to the laws of the State of California, and so on. Tarski's practice clearly indicates that he assumes the first sort of view when it comes to linguistic concepts and conventions of language.

Unfortunately there is nothing on the topic where one might expect to find it in [Kotarbiński, 1966]. Woleński says that such a view was "customary in the Warsaw School" [Woleński, 1989, 205], and in a letter to Neurath Tarski writes that "one must not forget that semantic expressions like designate, etc., have a relative character, that they must always be relativized to a particular language" [Tarski, 1992, 16–17] (my translation) and attributes this view to the Warsaw school generally. Though I have to confess to lacking clear enough intuitions about the individuation of "concepts" to settle the issue myself, this choice does seem to correspond to a good deal of ordinary thought and talk about illegality: though we recognize that the laws of New York and California differ in some respects, we treat the laws as settling the application conditions of a single notion of legality.

Now [Kotarbiński, 1966, 22] does contain a discussion of a closely related topic: indexicality or "occasionality". An aspect of this discussion bears mention now because it involves an unclarity that seems also to infect Tarski's usage. It bears mentioning that the discussion occurs in a chapter entitled "On Defectiveness in a Language"; that a language that expresses indexical or "occasional" concepts is "defective" ends up being one of the reasons that Tarski leaves relativization to a language in the mind of the reader. Here is the passage:

Among situations in which we readily fall victims to ambiguity, it seems advisable to mention those involving what is called *occasionality*, that is situations in which the meaning of a word varies according to the occasion. The words "I", "here", and "now" are striking illustrations of occasionality... The words "I" and "my", the meanings of which, to put it freely, have a certain personal shade, vary relatively as to their meaning accordingly to by whom they are used; the words "now" and "present" the meanings of which have a temporal shade, vary relatively as to their meaning according to when they are used... For instance, "I", used by Peter has such a meaning that it denotes Peter, and used by John has such a meaning that it denotes John...

This may give rise to misunderstandings, as in the case of the deceptive sign-board in a shop "Tomorrow on credit, today for cash". On reading that, as customer might be inclined to come on the next day so as to be able to buy goods on credit, but then the owner would point to the signboard and refuse to sell on credit. The text on the signboard preserves its form, but changes its meaning every day. Hence the trick resorted to by the shop owner is not honest: he makes consecutive statements which contradict one another, in doing which he avails himself of the fallacy that two sentences which have the same form always mean the same. But this alleged principle does not hold when it comes to sentences that include occasional expressions [Kotarbiński, 1966, 22–3].

First we can note that this, combined with Tarski's views about meta-mathematics according to which in "formalized languages ... the sense of every expression is uniquely determined by its form" [Tarski, 1983a, 165–6], it follows that there can be no "occasional" expressions in formalized languages, and hence taking metalinguistic terms like "sentence" or "consequence" or "truth" to be indexical for Tarski in his metalanguages is no good.[4]

More important for our purposes, though, is Kotarbiński's wavering between a more and a less sensible view of indexicality in the passage. The less sensible view is that when John and Bill both say "I am here", they utter sentences that *merely* share a form, but have completely distinct meanings as well as completely distinct denotations. This leaves out the obvious sense in which John and Bill both "say the same thing" with the sentence. Indeed, the most straightforward application of Kotarbiński's own account of indirect statement applied to indexicals would suggest that the conventions of English associate a single thought-content with all tokenings of the expression type "I": general conventions of language could not relate particular tokenings of "I" to their speakers. The more sensible view crops up in turns of phrase such as "the meanings of which have a certain personal (temporal) shade"; on this view, "I am here" has a conventional meaning of which Bill and John avail themselves to say something that involves different denotations.[5]

The passage strikes me a simply confused; Kotarbiński, if he had thought it through, could not have accepted the first view, but the first view is certainly more clearly present in the passage. Unfortunately, this waffling introduces a similar ambiguity into Tarski's conception. At [Tarski, 1983a, 263] Tarski writes that "the concept of truth depends, as regards both extension and content, on the language" and that "as soon as the discussion concerns more than one language, the expression ceases to be unambiguous". This makes it sound as though a linguistic term, applied to different formalized languages, changes not only its extension but also its content. But if this is really Tarski's view, then he really can't claim to show how to introduce an expression for "the" concept of truth into a whole family of metatheories, since the introduced expressions really would have nothing in common. However, something has gone wrong at [Tarski, 1983a, 263], since the passage says that the view that relativization to a language makes for a difference both in extension and in content was "already emphasized in the Introduction", but the relevant passage of the introduction, [Tarski, 1983a, 153], says nothing about content, but only that extension varies by language. Tarski never completely settled his usage of the terminology here, but what is clear is that when it comes to linguistic concepts and their relativization to a language, the "content of the concept" that Tarski tries to capture has to be something that is invariant across applications to different languages, something, as with the the treatment of consequence discussed above, is introduced into some theories by definition and into others axiomatically.[6]

Our reading will thus be that some concepts are essentially relative in the sense that they have extensions only relative to some further parameter and linguistic concepts are among such concepts. As noted above, we have an explicit statement to this effect in a roughly contemporaneous letter to Neurath. Frege might disapprove, but this is where the evidence points. Though Tarski talks in a few places about the "content" or "meaning" of a metalinguistic term being relative to the language to which it is applied, this talk cannot support his insistence that the implicitly relativized axiomatizations and definitions he gives always capture one and the same concept.

Tarski's aim in both the axiomatization of [Tarski, 1983g] and Definition 12 of CTFL is to introduce a term that expresses the one and only concept of being a closed well-formed formula, a concept over extensions for which the first generalizes, and which is granted an extension in the second case by the assumed relativization to the language of the calculus of classes as Tarski formulates it. For Tarski then, the concept applies only relative to a language. Since, as noted, being in a language isn't something for which the form of an expression can go proxy, Tarski counts on the reader to keep the relativization in mind while leaving it unstated in the metalanguage, since the metatheory is supposed to amount to a *mathematical* treatment of the object language. In this respect the relativization to a language is of the same status as the interpretation of the primitive terms of the metatheory—to the extent that the theorems don't settle it, it is something external to the formalized discipline but required to be known to the reader in order to understand the thoughts conveyed by theorems of the system.

Thus, once we account for Tarski's assumed views about the concepts in question, there is actually no departure here from Leśniewski at all. Tarski's aim in defining or axiomatizing an expression that expresses the concept of a sentence is to capture the one and only concept of a sentence. It is just that this concept, unlike the concept expressed by Leśniewski's "ε", gets an extension only relative to a language, and this relativization must be kept in the mind of the one who "practises" the deductive metatheory for the object language. When so relativized, the intended concept is expressed as long as certain characteristic claims are stated by theorems of the metatheory. The metatheory might as well then be explicitly relativized to the object language, but since this would bring the problematic notion of an expression being "in a language", and thus the concomitant views about conventions, onto the page for explicit treatment, Tarski forces the relativization to a language into the mind of the reader.

We can thus say the same for the concept of consequence as Tarski construes it in the works under consideration: there is a single concept of one sentence's following from others which determines an extension relative to a set of rules of proof; the aim of both a general axiomatic theory of this relation and any definition relativized to some particular deductive science is to introduce an expression properly constrained by the axioms of the metatheory to express this concept, with the relativization to the object language, or some object language or other, left in the mind of the reader. The aim is, as elsewhere in metamathematics construed in the Intuitionistic Formalist way, to introduce an expression that expresses an intuitive concept but is itself defined entirely by terms from "the morphology of language".

2.2 Explicit definition

I turn now to "Some Methodological Investigations on the Definability of Concepts" (DC), the paper Tarski published in the proceedings of the Prague *Vorkonferenz* for the Unity of Science Congress to be held in Paris in 1935. Though the paper was delivered in 1934 and published in 1934 (Polish) and 1935 (German), it contains material that dates from 1926 (§ 1) and 1932 (§ 2) [Tarski, 1983k, 297]. Since it was delivered well after Tarski had developed his semantic techniques the fact that he was willing to present and publish it at those late dates shows that as late as the 1934 *Vorkonferenz* Tarski was still a committed Intuitionistic Formalist, despite the fact that semantics has a (now) obvious application to the topics discussed.

On my reading Tarski conceived of the work as a study of the conditions under which a theory achieves the Intuitionistic Formalist's goal of setting out a theory that captures intuitive meaning as effectively as possible. The paper is widely neglected in part because it treats some standard topics in a way that involves significant idiosyncrasy and because, as Hodges notes, "The mathematical interest is almost trivial; certainly an ingenious and ambitious mathematician like Tarski would never have published it as a contribution to mathematics" [Hodges, 2008, 108]. However, if we view the paper as a contribution to the Intuitionistic Formalist project of expressing concepts by capturing their analyses within a deductive theory, the otherwise strange modifications Tarski makes to Padoa's method and to the notion of categoricity make good sense. It will also be of use to us to study the paper because it contains Tarski's most extended discussion of explicit definition.

2.2.1 Defining definition

Tarski defines "definition" and explicitly ties its significance to the the assertibility of certain sentences on the basis of a deductive theory in this passage:

Every sentence of the form:

$$(x) : x = a . \equiv . \phi(x; b', b'', \dots),$$

where '$\phi(x; b', b'', \dots)$' stands for any sentential function which contains 'x' as the only real variable, and in which no extra-logical constants other than 'b''', 'b'''', ... of the set B occur, will be called a *possible definition* or simply *a definition of the term 'a' by means of the terms of the set B*. We shall say that the term 'a' *is definable by means of the terms of the set B on the basis of the set X of sentences*, if 'a' and all terms of B occur in the sentences of the set X and if at the same time at least one possible definition of the term 'a' by means of the terms of B is derivable from the sentences of X [Tarski, 1983k, 299].

It is not difficult to see why the concept of definability, as well as all derived concepts, must be related to a set of sentences: there is no sense in discussing whether a term can be defined by means of other terms before the meaning of those terms has been established, and on the basis of a deductive theory we can establish the meaning of a term which has not previously been defined only by describing the sentences in which the term occurs and which we accept as true [Tarski, 1983k, 299, note 1].

(The curious "possible definition" can be be found at [Kotarbiński, 1966, 198].) The note expresses the paper's commitment to Intuitionistic Formalism. The Intuitionistic Formalist is interested in associating intuitive concepts with the expressions of a theory. Now which intuitive concepts can be associated with the terms of a theory depends on which sentences of the language of the theory are accepted as true within the theory. The significance of an explicit definition thus lies in the demonstration that the axioms, plus the formation and transformation rules, fully individuate the intuitive interpretation of a term relative to the interpretations of other primitive non-logical terms.

Now it is in this moment that an important feature of Intuitionistic Formalism comes to light. The entire article is concerned with "the definability of concepts". This means, I take it, that a definition determines the content of the the concept expressed by its *definiendum*, and the only possible determination could be identity: a term defined by another (simple or complex) shares its intuitive meaning with it. What should strike us here is that it follows that the intuitive meanings of terms aren't determined compositionally: a complex symbol's meaning isn't entirely determined by the meanings of its parts and how they are syntactically put together, it depends also on the intuitive meanings of other symbols for which possible definitions equating them and it are provable in the theory. We see here how very different the Intuitionistic Formalist account of meaning is from the referential semantics to which Tarski's own semantic work gave rise. Furthermore, as much as Leśniewski himself demanded a compositional account in his early work [Leśniewski, 1992b], in Tarski's hands Intuitionistic Formalism is incompatible with it because the intuitive content assigned to an expression depends on the set of theorems of the overall deductive theory, and not merely on the intuitive contents of its parts—indeed, all expressions get their intuitive content assigned "top down" by looking at the set of theorems in which they figure.

What the provability of a definition shows is that *definiens* and *definiendum* share their meaning to the extent that either of their meanings is determined by the theory. To the extent that the theory fails to determine this, the intuitive content expressed by both symbols is left open. It is because of this sort of possibility that Tarski introduces the concept of a theory's being complete with respect to its specific terms in § 2 of the article, for in a theory that meets this condition the intuitive meanings of all terms are uniquely individuated.

2.2.2 Two conceptions of definition

An important thing to recognize about Tarski's conception of explicit definition is that he stands within a minority tradition in 20th century thought about what an explicit definition is. Call the two views of explicit definition at issue the *abbreviative* conception of definition and the *theory-relative* conception of definition. A definition states some sort of equivalence between a *definiendum d* and its *definiens d'*. On the abbreviative conception, this equivalence is a matter of stipulated equivalence and definition is thereby really a metalinguistic affair. On the theory-relative conception, on the other hand, a definition is simply an equivalence that follows from a theory. (See [Belnap, 1993] for a clear recent treatment of the latter conception.) Unlike the abbreviative conception, the theory-relative conception holds that there is no difference between the '$\leftrightarrow$' of ordinary mutual implication of theses in a theory and the '$=_{df}$' of definition. That this distinction is so often made in formal and informal presentation is a symptom of how entrenched the abbreviative conception is. Tarski, by contrast, got his start in publication with an innovation that made it possible to get rid of this distinction and thus base a system of the sentential calculus on a single connective [Tarski, 1983j], something Leśniewski had wanted but was unable to supply himself [Leśniewski, 1992h, 418ff].

The contrast between these two conceptions was a topic for discussion among Tarski and his teachers (cf. [Woleński, 1989, 100, 107], [Hodges, 2008, 101–2]).[7] Both Kotarbiński and Leśniewski discuss the topic and Tarski briefly notes it at [Tarski, 1983e, 166].[8] Kotarbiński sides with the stipulative conception. Early in the *Elementy* he distinguishes "axiomatic pseudo-definitions" in which both *definiendum* and *definiens* are used—that is definitions taken as the theory-relative conception takes them, as ordinary biconditionals—and maintains that axiomatic pseudo-definitions are not definitions "in the strict sense of the term" [Kotarbiński, 1966, 27]. In a later section on definitions in "The Deductive Method", he writes:

> Definitions include such formulas as "means the same as" or "is equivalent to", and also names of the symbols being defined … defintions cannot themselves belong to the sentential calculus; the same holds for other definitions of many deductive systems … For every proper definition, we can choose … a corresponding pseudo-definition as belongs to the system, whereas the proper definition itself remains

outside the system as a not indispensable comment concerning the sameness of meaning of some two inscriptions ... they provide information that a given symbol may be used to replace another given symbol in the process of inference ... In purely formal transformations a formulation of the kind "*A* is a graphical substitute of *B*" would suffice. Such definitions might be called *graphical*. But usually we have to do with semantic definitions formulated by means of such phrases as "means the same as" ... Such definitions, apart from providing information about permissible substitutions, also provide the information that he who formulates them associates the same meaning with the *definiendum* as with the *definiens*. This is important in communication, if the listener or reader is to understand the sentences of the system in the same sense as does their author or expounder [Kotarbiński, 1966, 244–5].[9]

(Note here that the psychologisitc conception of meaning that informs Intuitionistic Formalism is clearly in evidence in the penultimate sentence.) Though Kotarbiński here stresses the communicative utility of abbreviative, metatheoretic definitions, he admits their dispensability and notes the dispensability of the notion of meaning when it comes to "formal" transformations. What Tarski saw more clearly here, in line with Intuitionistic Formalism as he understood it, was that "axiomatic" definitions simply record in fact what the other sort merely purport to do in word: express the intersubstitutability of two expressions.

Though Leśniewski's published work contains no explicit discussion of the two sorts of definition, Leśniewski was aware of the difference at least to the extent that his position changed between his early and late work. Early on he maintains that

> Inadequacies of verbal representation ... may, however, hinder research when, e.g., inadequate symbols of certain contents are mistakenly regarded as adequate. In such cases whole series of fundamentally false theories may grow out of inadequate representation; such theories can be exemplified by, say, the frequent opinion that definitions are analytic propositions about the objects represented by the subject ... This opinion ... [is] based on the fact that instead of formulating a proposition about the expression we are going to define, we formulate a proposition about the object of which the expression in question can only be a symbol [Leśniewski, 1992b].

Later, however, he came around to a view shared with Tarski that Kotarbiński's "axiomatic pseudo-definitions" are the only real definitions:

In defining the functions of the theory of deduction in terms of other such functions, both Scheffer and Nicod use a special equal-sign for definitions which they do not define in terms of the primitive functions of the system ... This circumstance makes it difficult to say whether Nicod's theory of deduction is in fact constructed out of the single primitive term '|'.

In 1921 I realized that a system of the theory of deduction containing definitions would actually be constructed from a single term only if the definitions were written down with just that primitive term and without recourse to a special equal-sign for definitions [Leśniewski, 1992e, 418].

Tarski sides with Leśniewski in his later period here, holding that a definition is simply an equivalence provable from a theory.

Another restriction on Tarski's conception must be noted: that an expression is everywhere substitutable for another preserving theoremhood if and only if an explicit definition is provable from the same theory depends on specific features of the assumed logic. For a toy example, suppose my theory consists of the two axioms Fa and Fb and the logic is the null logic in which nothing follows from anything. Here a and b are everywhere substitutable preserving theoremhood, but obviously no explicit definition is provable from the theory. Tarski's conception assumes the analogue of Beth's definability theorem for STT [Vaught, 1986, 864], and it would make the same demands of any other system to which it was applied.

2.2.3 Padoa's method

The explicit aim of § 1 of "Some Methodological Investigations on the Definability of Concepts" is somehow to support the applicaibility of Padoa's method for establishing that none of the primitive terms of a deductive theory are mutually definable:

Many years ago A. Padoa sketched a method which enables us to establish, in particular cases, the undefinability of a term by means of other terms. In order, by this method, to show that a term 'a' cannot be defined by means of the terms of a set B on the basis of a set X of sentences, it suffices to give two interpretations ("*Interpretationen*") of all extra-logical constants which occur in the sentences of X, such that (1) in both interpretations ("*Interpretationen*") all sentences of the set X are satisfied ("*erfüllt*") and (2) in both interpretations ("*Interpretationen*") all terms of the set B are given the same sense ("*Sinn*"), but (3) the sense ("*Sinn*") of the term 'a' undergoes a change. We shall here present some results which provide a theoretical justification for the method of Padoa and, apart from this, appear to throw an interesting light on the problem of definability [Tarski, 1983k, 299–300] [Tarski, 1986b, 640].

The quotation marks are in the German text but are omitted by Woodger. (Leśniewski also scare-quotes "interpretation" while discussing "the well known method of 'interpretation'" at [Leśniewski, 1992h, 261].) Woodger's omission of them is significant: what Tarski proposes to do in § 1 is to tell us how to remove the scare quotes; that is, he intends to replace Padoa's appeal to "interpretation" with something supposedly more rigorous.

Given the position of the paper in *Logic, Semantics, Metamathematics*— after CTFL and ODS—and its publication in 1934, one might expect that Tarski would get rid of "sense", "interpretation" and "satisfaction" in favor, in the obvious way, of his semantics. But this is not what he does (cf. [Hodges, 2008, 125]) and the appearance in 1934 of material not updated since 1926 shows just what the state of Tarski's thinking about his semantics and its relation to meaning was in 1934.[10] We will return to what Tarski could have done in 1934, and what the fact that he didn't do it shows, below. Here the point is to describe what Tarski does do.

Though it is tempting to think of Padoa as some sort of proto-model-theorist, as already noted there is nothing semantic about [Padoa, 1967].[11] The passage in which Padoa presents his "method" for proving that no primitive term of a deductive theory can be defined in terms of others reads thus:

Let us assume that after an *interpretation* of the system of undefined symbols that verifies the system of unproved propositions has been determined, all these propositions are still verified if we suitably change the meaning of the undefined symbol *x* only. Then, since the meaning of *x* is not *individualized* once we have chosen an *interpretation* of the *other* undefined symbols, we can assert that it is impossible to deduce a relation of the form *x* = *a*, where *a* is a sequence of other undefined symbols, from the unproven propositions.

Conversely, in order to be able to assert that it is impossible to deduce, from the unproved propositions, a relation of the form just mentioned we must show that the meaning of *x* is not individualized once we have chosen an interpretation of the system of the other undefined symbols; and this we do by establishing an interpretation of the system of undefined symbols that verifies the system of unproved propositions and that still does so if we suitably change the meaning of x only. [Padoa, 1967, 122].

This might sound semantic until we take note of Padoa's conception of interpretation, meaning and related notions, which, as we have seen (§ 1.4.1) [Padoa, 1967, 120–21] is completely psychologistic: Padoa's conception of interpretation exactly matches the Intuitionistic Formalist's idea of associating an intuitive meaning with a symbol.[12]

Viewed in this light, the interest of Padoa's method is that it concerns the extent to which the formal role of a symbol within a theory is constrained by the theory down to the point where, given the interpretations of the other symbols, a rational thinker who accepts the theory can associate only one particular intuitive concept with a symbol. This is what captures Tarski's attention in Padoa's method. Tarski's interest is in understanding the extent to which a theory adequately achieves the Intuitionistic Formalist ideal of bearing a unique intuitive interpretation as determinately as possible.

Now [Tarski, 1983k] is a faithful rendering of Padoa's approach understood *as Padoa understood it*. What Tarski does is to replace Padoa's talk about whether or not the "meaning" of a symbol is "individualized" by a mental choice of meanings for other symbols by a deductive condition: that, for a term a and a set of other primitive terms B, it be the case that if we introduce another term a' and restatements of all axioms involving a by their substitution variants with a', $a = a'$ be provable from the extended theory. When this condition obtains, one cannot associate two different meanings with a and a' while taking the theorems of the extended theory to be true (assuming that the theory is extensional). The more general statements of the conception come as Theorems 2 and 3, in which definability and indefinability are conceived of as involving generalizations on the provability or (syntactic) consistency of the relevant identities or their negations [Tarski, 1983k, 303–4].

What in Padoa's presentation goes on in one's mind while contemplating the deductive theory, in Tarski's refinement goes on deductively within the system itself; it is for this reason that Tarski holds that he has provided "the proper theoretical foundation for the method of Padoa" [Tarski, 1983k, 305]. The idea of associating meanings with symbols is still present in the article, since the interest in the conditions stated in Theorems 2 and 3 concerns the possibility of associating different meanings with symbols of the theory, but Padoa's idea that we will find, when we contemplate the system, that we either can or cannot "vary the meaning" we associate with a symbol "in the mind" while still taking the primitive sentences of the theory to be "verified" has been replaced by conditions on derivability involving identity statements formulated with the symbols.[13] This is exactly in line with "metamathematics" as Tarski construes it. If definability and eliminability of primitives is to be studied metamathematically, the issues have to be cast in a form where what they concern is the form of expressions. This, then, demands that Padoa's mental variation of meanings by replaced by the introduction of two symbols differing in form, each of which bears one of these meanings.

2.3 Categoricity and completeness of terms

2.3.1 Provable monotransformability

We turn now to the second section of "Definability of Concepts", representing work that was complete by June, 1932. One might expect, given the section's focus on categoricity and the late date, that it would be grouped with Tarski's semantic works, but there is nothing semantic about it, either: it is, rather, a further contribution to to Intuitionistic Formalism. Indeed, I think it ranks as Tarski's deepest attempt to contribute to the basic Intuitionistic Formalist project as he understood it, for in the section Tarski attempts to establish some general results about the conditions under which a deductive theory: (a) completely determines the concepts expressed by its non-logical vocabulary, and (b) expresses *all* of the intuitive concepts of the domain from which these concepts are drawn—e.g., the conditions under which a theory expresses, for instance, all geometrical concepts. The second point of course involves a view about what makes a concept a *geometrical* concept, etc. and we will examine what Tarski does with this. Full determination of the concepts expressed by the sentences held true in a theory was the Intuitionistic Formalist ideal, and in this section Tarski gives criteria for the conditions under which it has been realized.

Tarski first introduces the notion of a set of sentences being "essentially richer" than another "with respect to specific terms":

> The problem of the *completeness of concepts*, to which we now turn, is also closely related to certain problems concerning systems of sentences, and indeed to the problems of completeness and categoricity, although the analogy does not extend so far as in the previous case.
>
> In order to make the problem of completeness precise, we first introduce an auxiliary concept. Let X and Y be any two sets of sentences. We shall say that the set Y is *essentially richer than the set X with respect to specific terms*, if (1) every sentence of the set X also belongs to the

set Y (and therefore every specific term of X also occurs in Y) and if (2) in the sentences of Y there occur specific terms which are absent from the sentences of X and cannot be defined, even on the basis of a set Y, exclusively by means of those terms which occur in X. [Tarski, 1983k, 308].[14]

As Tarski notes next, for any set of sentences there is trivially an essentially richer set: just add a logical truth with a term that doesn't occur in the original set. In order to devise a condition of some interest here, Tarski makes use of the notion of categoricity, but with peculiar twists. He introduces it as something familiar: "as is well known, a set of sentences is called categorical if any two interpretations (*"Interpretationen"*) (realizations (*"Realisierungen"*)) of this set are isomorphic" [Tarski, 1983k, 309]. (Once again the quotation marks are in the original but have been stripped by Woodger.) Tarski's footnote to Veblen encourages the thought that genuine semantics is now at issue in § 2 but it turns out immediately that Tarski is interested entirely in a deductive conception of categoricity:

Consider now any finite set Y of sentences; 'a', 'b', 'c', ... are all specific terms which occur in the sentences of Y, and '$\psi(a,b,c,...)$' is the conjunction of all these sentences. The set Y is called *categorical* (or *provably categorical*) if the formula

(V) $(x',x'',y',y'',z',z'',...) : \psi(x',y',z',...) . \psi(x'',y'',z'',...).$

$$\supset .(\exists R).(R\tfrac{x',y',z',...}{x'',y'',z'',...})$$

is logically provable [Tarski, 1983k, 310].

Here the R-formula gets the following interpretation:

Let us say that the formula

$$R\frac{x',y',z',...}{x'',y'',z'',...}$$

is to have the same meaning as the conjunction ... in words: 'R *is a one-one mapping of the class of all individuals onto itself, by which* x', y', z', ... *are mapped onto* x'', y'', z'', ..., *respectively* [Tarski, 1983k, 310].

Rather than thinking of categoricity in terms of its being a theorem of a metatheory for the deductive theory at issue that any two realizations of its axioms are isomorphic whether this can be proved in the theory or not, categoricity is here construed as its being provable in the background logic of the theory (simplified type theory, as usual) that for any two ordered n-tuples of things (of any type; as Tarski notes the notation here is really schematic over the type-theoretic hierarchy [Tarski, 1983k, 310 note 1]) if they pairwise have the features attributed to them by the axioms, then there is a 1–1 mapping of the set of all individuals onto itself that maps them onto one another.[15]

Categoricity is important to Tarski because he holds that:

> On various grounds, which will not be entered into further, great importance is ascribed to categoricity. A non-categorical set of sentences (especially if it is used as an axiom system of a deductive theory) does not give the impression of a closed and organic unity and does not seem to determine precisely the meaning of the concepts contained in it. We shall therefore subject the original definition of the concept of completeness to the following modification: a set X of sentences is said to be *complete with respect to its specific terms* if it is impossible to construct a categorical set Y of sentences which is essentially richer than X with respect to specific terms. In order to establish the incompleteness of a set of sentences it is from now onwards requisite to construct a set of sentences which is not only essentially richer but also categorical. Trivial constructions in which the meaning of the newly introduced specific terms is quite indeterminate are thus excluded from the beginning [Tarski, 1983k, 311] [Tarski, 1986b, 647].

A categorical axiom system, then, "determines precisely" the meanings of the terms it contains. (I take it that Tarski's "concepts" in the second sentence is to be ignored in favor of "terms" from the last, since it is only the latter that are actually "contained" in an axiom system.) This is what an Intuitionistic Formalist wants: that a deductive theory constrain the concepts expressed by its terms down to uniqueness. Completeness is then an additional matter, corresponding to (b) above: a deductive theory is complete with respect to its specific terms if one cannot extend it to contain a new term the meaning of which is also completely determined.

There are two modifications to the familiar conception of categoricity here: provability in the assumed logic and the requirement of a 1–1 mapping from the set of all individuals onto itself. Contrast the first with the standard model-theoretic conception of categoricity. If a model of a set of sentences is a set of assignments of values to variables, denotations to names and so on, and a domain for the quantifiers that satisfies all of the sentences, then a set of sentences is categorical just in case every model is isomorphic. With that notion of a model on board, anyone with even a passing acquaintance with the notion of categoricity would define it in no other way. Compare here Veblen's remarks on the categoricity of his axioms for geometry:

Inasmuch as the terms *point* and *order* are undefined, one has a right, in thinking of the propositions, to apply the terms in connection with any class of objects of which the axioms are valid propositions. It is part of our purpose however to show that there is *essentially only one* class of which the twelve axioms are valid. In more exact language, any two classes *K* and *K'* of objects that satisfy the twelve axioms are capable of a one-to-one correspondence such that if any three elements *A*, *B*, *C* of *K* are in the order *ABC*, the corresponding elements of *K'* are also in the order *ABC*. Consequently any proposition which can be made in terms of points and order either is in contradiction with our axioms or is equally true of all classes that verify our axioms [Veblen, 1904, 346].

The understanding in Veblen is fully semantic: there is nothing here that of itself would restrict application of the notion of categoricity to finite sets of sentences, as Tarski's conception does, and there is also nothing about it being a logical truth of the logic of the language in which the axioms are stated that any two sets of things of which they hold are such that there is a 1–1 mapping of *V* onto itself that 1–1 maps them onto one another. Tarski's focus, by contrast, is on construing categoricity in terms of what can be derived from what. Notice on this score that the parenthetical *"(or provably categorical)"* of the English does not appear in the German of the article, but is added to the 1956 English translation [Tarski, 1986b, 650]; by 1956 Tarski wanted it known that he was aware that the earlier conception had been bent to his 1932 purposes. Tarski also notes at the end of the translation that the remarks on categoricity do not reflect his views at the time of the translation.

What does this add up to? Tarski is trying to make categoricity a matter of what is provable *in* the language of a theory, rather than something that is provable *about* a theory. This is entirely in line with metamathematics as he construes it in the early 1930s: just as Padoa's method receives its "theoretical justification" by being recast as a deductive condition concerning two symbols rather than an intuitive condition about the variability of the meaning one mentally attaches to a single symbol, so too the notion of categoricity isn't construed in terms of model theory or in terms of mentally associated concepts, but is rather being conceived of as a deductive condition. The sense in which a non-categorical set of axioms doesn't fully determine the concepts expressed by the specific terms it employs is explicated in terms of the extent to which the intuitive concepts that could be expressed by terms are constrained by the axioms being held true.

Now Tarski's presentation distinguishes categoricity from what he calls "monotransformability", where a monotransformable set of sentences is such that one can prove:

$$(VI) \ (x',x'',y',y'',z',z'', \ldots, R',R'') : \psi(x',y',z', \ldots) \ . \ \psi(x'',y'',z'', \ldots).$$

$$\supset \ . \ (R' \tfrac{x',y',z',\ldots}{x'',y'',z'',\ldots}) \ . \ (R'' \tfrac{x',y',z',\ldots}{x'',y'',z'',\ldots}) \ . \ \supset R' = R'' \ \text{[Tarski, 1983k, 313]}$$

That is, a monotransformable set of sentences is one that is categorical and where, in addition, there is only one 1–1 mapping of the set of individuals onto itself compatible with the truth of the theory. Monotransformability inherits the requirement of "provability" in Tarski's presentation, so we are again talking about a deductive condition here. Tarski holds that monotransformability is sufficient for the completeness of a theory with respect to its specific terms in Theorem IV [Tarski, 1983k, 314]. The converse is left open and, as Vaught notes [Vaught, 1986, 874], it may still be open, though as we will see this is surely due to the lack of any logical handle on "competeness with respect to specific terms" other than monotransformability. This makes clear again that the meaning in the article cannot be equated with extension, for if meaning were extension then completeness with respect to specific terms and mono-transformability would be transparently one and the same property. The meaning at issue when Tarski holds that provable monotransformability is sufficient for complete determination of the meaning of a theory's terms is intuitive meaning. For Tarski's purposes, however, the distinction between categorical and non-categorical sentences matters little; in both cases one could exchange the ideas associated with the specific terms of the theory while still taking the theory to be true. Monotransformability is the important property, since a monotransformable theory constrains (the contents of) the concepts associated with its specific terms down to uniqueness.

Construing provable categoricity as Intuitionistic Formalism demands, consider oneself, in the position of the Leśniewskian mathematician trying to associate intuitive concepts with the terms of a theory. Following Tarski's examples, consider a theory for one-dimensional descriptive or metric geometry; the first has "x is between y and z" as a primitive; the second adds "the segment xy is congruent to the segment zw". Now calling to mind one's idea of the real line, with points 0 and 1 distinguished, one intuitively distinguishes, e.g., the segment with endpoints 0 and 1, and the segment with endpoints 1 and 2. But one also sees that any defined term of one-dimensional metric geometry with which one might try to associate one's idea of the segment with endpoints 0 and 1 could just as well have one's idea of the segment with endpoints 1 and 2 associated with it: nothing in the theory itself demands one interpretation as opposed to the other, even though the theory is categorical. There are non-identity automorphisms, and hence different sets of intuitive concepts can be associated with the terms of theory: one can associate the concept of the interval from 0 to 1 with "01" or one can associate the concept of the interval from 1 to 2 with it. If a theory is provably categorical (or not categorical at all) but provably not monotransformable, then one knows that there is more than one way to associate intuitive meanings with its specific terms compatible with taking it to be true. If the theory is monotransformable, then, since the identity automorphism is available and one can prove (VI), one knows that the concepts associated with the terms of the theory are constrained down to uniqueness.

Now since the theory is extensional, it won't determine intuitive meanings past their being individuated extensionally. This might seem a difficulty on the grounds that there are obviously expressions (e.g. "Hesperus" and "Phosphorus") that intuitively differ in meaning but are extensionally equivalent. But Tarski would reject the example: any two terms whose equivalence (in the case of names, an identity statement) is provable in the theory have to be attributed the same intuitive meaning, since Tarski's conception of definitions doesn't distinguish them from other sorts of equivalences, and definitions define "concepts".[16] Tarski's wording certainly indicates that he holds that monotransformability is sufficient for determining intuitive meanings *completely*; witness again his comment that "A non-categorical set of sentences ... does not give the impression of a closed and organic unity, and does not seem to determine precisely the meaning of the concepts contained in it" [Tarski, 1983k, 311]. This would seem to indicate that a categorical, and certainly a monotransformable set of sentences *does* determine the intuitive meanings of its terms completely. But since the theory is extensional, this means that intuitive meanings are individuated at best extensionally.[17]

Unfortunately, at this point in §2 of DC it becomes clear that Tarski, searching for mathematical expressions of his views about concepts, has landed himself in a position that simply isn't consistent with the views that animate the other papers: concepts according to Intuitionistic Formalism aren't extensionally individuated. This is an obvious consequence of Leśniewski and Kotarbiński's views about connotation and denotation. In particular the major goal of CTFL, to express the conception of truth expressed in the "semantical definition" of §1, would be lost if concepts and their contents were individuated extensionally, since, as we will see (§ 4.3.1), Tarski recognizes that definitions that clearly aren't intuitively adequate can still be extensionally adequate. For this reason I will set § 2 of DC aside for the remainder of this work as something of an experiment that went wrong. The problem, one can see, was waiting for Tarski all along: as a logician he was stridently extensionalist, while as a philosopher he accepted an account of concepts on which they weren't individuated extensionally. The problems of § 2 of DC show this tension finally coming to light when Tarski sets his mind to rigorously working out certain theses about concepts in a deductive theory. The tension, it seems, was never really resolved; it ceased to be important when Tarski left Intuitionistic Formalism behind in 1935.

2.3.2 Absolute monotransformability

That said, § 2 is still interesting in itself, and so we will continue our examination of it. A provably monotransformable theory uniquely individuates the intuitive concepts expressed by its specific terms and as such perfectly meets the Intuitionistic Formalist ideal of capturing the contents of our concepts within the deductive structure of an axiomatic theory. We turn now to the other modification of the usual notion of categoricity in the article. Tarski comments on it in a note:

> We use the word 'categorical' in a different, somewhat stronger sense than is customary: usually it is required of the relation R which occurs in (V) only that it maps x', y', z', ... onto x'', y'', z'', ... respectively, but not that it maps the class of all individuals onto itself. The sets of sentences which are categorical in the customary sense can be called *intrinsically categorical*, those in the new sense *absolutely categorical*. The axiom systems of various deductive theories are for the most part intrinsically but not absolutely categorical. It is, however, easy to make them absolutely categorical. It suffices, for example, to add a single sentence to the axiom system of geometry which asserts that every individual is a a point (or more generally one which determines the number of individuals which are not points) [Tarski, 1983k, 310 note 2].[18]

Intrinsic categoricity isn't sufficient for absolute categoricity because if the set of all individuals and the set of individuals to which the theory is committed are infinite then we can describe two isomorphic models that leave the individuals to the existence of which the theory isn't committed out of 1–1 correlation. To take a simple example, if my theory is "there are infinitely many individuals" and the set of all individuals is the natural numbers, then the even naturals and the naturals $\neq 3$ are both models of my theory, but the odd naturals and 3 are not in 1–1 correspondence. Hence facts about infinite sets determine that there are theories all models of which are isomorphic which are not such that for any two models there is a 1–1 correlation of the set of all individuals that also 1–1 correlates *those* two models: one cannot 1–1 correlate the even naturals and the naturals $\neq 3$ while also 1–1 correlating the set of all naturals with itself.[19]

Why does Tarski impose this additional requirement? And why does he impose it when a theory can be trivially but extraneously modifed to satisfy it? This is where the second aspect of "completeness with respect to specific terms come in". As Tarski understands a complete theory in this sense, it is complete both in that it individuates the intuitive meanings of its specific terms uniquely, *and* in that it doesn't allow any additional terms not definable in terms of the old ones to be added. Here is the explanation:

> If we wish to use the words 'categorical' and 'monotransformable' in the intrinsic sense ... then Th. 4 in the above form would be false. In order that it should remain valid the definition of completeness must be weakened, *absolute completeness* must be replaced by *intrinsic completeness* or by *completeness with respect to intrinsic concepts*. For this purpose it would be necessary to give first an explanation of what is meant by 'intrinsic concept'. All this would complicate the exposition considerably [Tarski, 1983k, 314].

The problem is that if the theories in question didn't exhaust the set of all individuals, a monotransformable theory would be extendable by a (provably) categorical theory that contained a term not definable by the original theory's terms in the extended theory. As Tarski notes, this need for an otherwise extraneous recasting of the theories at issue could be avoided if "intrinsic concept" could be defined: intrinsically monomorphic theories would completely determine their "instrinsic" concepts and the existence of categorical extensions would be no issue. But Tarski shies away from committing himself to such an account of being "intrinsic". On the other hand, if a theory is absolutely monotransformable then no categorical extension can introduce a term that has a determinate meaning but for which a possible definition in terms of the old vocabulary isn't provable: since an absolutely monotransformable theory determines the whole of V up to the identity automorphism, there is no object left over in the STT hierarchy over V for any new term to have as its extension. This being so, by the result of § 1 (Beth definability for STT), a possible definition will be provable that reduces the new theory to a definitional extension of the old one.[20]

Tarski refers the reader to the end of 1935's "On the Limitations of the Means of Expression of Deductive Theories" (which we learn at [Tarski, 1983i, 384 n. 1] dates to the same period as § 2 of [Tarski, 1983k] and "belongs to the same circle of ideas" as it) where he writes, of the application of Theorem 1 of the article (which states that that which is logically definable is invariant under 1–1 permutations of the class of all individuals) to particular deductive theories:

> In conclusion we note the following: when applying Th. 1 to special deductive theories and to general metamathematics we must restrict ourselves to those axiom systems from which it follows that there are no individuals outside the domain of discourse of the theory discussed. We could remove the restriction by means of an appropriate generalization of Th. 1. To obtain this generalization we should introduce a new notion, in fact the notion of a *sentential function intrinsic under a given variable (or constant)* '*a*' or, in other words, the notion of a *property intrinsic for a given class a*. Roughly speaking, a property is intrinsic for a class *a* if it involves exclusively elements of *a*, subsets of *a*, relations between elements of *a*, etc. (and not, for example, individuals outside of *a*) [Tarski, 1983i, 392].

The explanation of "intrinsic" in the last sentence of the quote is a 1956 addition to the English translation that does not appear in the 1935 German version of the paper. That runs, rather:

> In conclusion we note the following: when applying Th. 1 to special deductive theories and to general mathematics we must restrict ourselves to those axiom systems from which it follows that there are no individuals outside of the "domain" of the theory considered. We could remove the restriction by means of an appropriate generalization of Th. 1 and by considering only sentential functions that are—in a certain sense that we will not here further make precise—"inner" in relation to the primitive terms of the contemplated theory [Tarski, 1986h, 212] (my translation of the alternate material).

Tarski appears to be attempting to cover his tracks in the 1956 English version by supplying a non-mysterious semantic understanding of "intrinsic". While the 1956 English version offers a rather clear model-theoretic understanding of an "intrinsic" concept relative to a sub-domain of the set of all individuals (compare here the remarks on truth in an individual domain at [Tarski, 1983a, 199ff]) through a straightforward relativization that makes no attempt to capture why some relativizations might be "intrinsic" relative to a theory while others would not, the 1935 German version holds out the hope of capturing some sense of "intrinsic" that is of philosophical interest. This passage also does away with any lingering doubt that the conception of meaning in DC is psychologistic rather than semantic: the 1956 modification is so obvious—and, indeed, is closely related to the discussion truth in a domain at [Tarski, 1983a, 199, 239]—that if Tarski had intended it in 1932-5 he would simply have stated it. Since he doesn't, there must be some reason, and my suggestion is that the reason is that he is still thinking of "interpretation" in Intuitionistic Formalist terms. Presumably, given the discussion in the article of arithmetic, geometry and mechanics, "intrinsic" concepts of, say, geometry would be all and only the *geometrical* concepts. But doing things this way would commit Tarski to give an account of what makes a concept a geometrical concept, and it is this philosophical heavy weather that he shies away from, preferring instead to formulate things in terms of absolute categoricity.

Thus the modification of "categorical" to its absolute sense is a fudge that allows Tarski to raise an issue he doesn't take himself to be able to settle. The effect of the move is to take the selection of the set of intuitive concepts we are trying to capture off the table and to put it into the mind of the mathematician in the assumed suitable selection of V for the application of STT. So, for instance, spotting ourselves the claim that our concepts of various properties of the real line hang together as some sort of philosophically important whole, making V the set of points on the real line then allows us to state it as a theorem that metric geometry plus the introduction of terms for 0 and 1 results in a theory that: (a) uniquely determines the concept expressed by every term of the arithmetic of the real numbers, and (b) exhausts the supply of concepts of the properties of the real line. Tarski would like to free himself of the need for the fudge factor, but lacking a way to do so he has to push the issue back into the assumptions held in mind by the mathematician.

So § 2 of DC is an attempt to state some mathematical conclusions about what has to be provable in a theory for it to meet the Intuitionistic Formalist goal of building a theory that allows us to express all of our thoughts about a domain in such a way that the structure of the theory uniquely settles which concepts are expressed by which terms, and which thoughts by which sentences. One can see from notes like [Tarski, 1983k, 319], Tarski's interpolation of "or provably categorical" in the 1956 translation, as well as from the related 1956 modification to the treatment of "instrinsic" in [Tarski, 1986h] that Tarski was later somewhat chagrined about the lengths to which he had gone in modifying the mathematics to suit philosophical purposes in the article. As I have noted, substantial aspects of the section are inconsistent with the rest of Tarski's views and for this reason I will leave the section out of the account when interpreting the other works.

However § 2 does tell us a good deal about Tarski's interests as late as 1934: he was still a fully committed Intuitionistic Formalist. His interests in definability and categoricty had a strong Intuitionistic Formalist bent; both were significant not model-theoretically but for their role in formally constraining the expression of intuitive concepts within a deductive theory. His failure to use his semantics to work out Padoa's method or to treat categoricity and definability model-theoretically shows, likewise, that even into 1934 Tarski was still squarely within the earlier paradigm, thinking of deductive relationships within a theory in terms of primitively valid transformation rules, and thinking of the meaning of a sign as the content of a mental state in a way that would have been familiar to the Twardowski of 40 years earlier [Twardowski, 1977, 8–10].[21]

2.4   Theory and concept

Tarski's treatment of definability in [Tarski, 1983k] is focused on the way in which the structure of a deductive theory constrains the concepts its terms can be taken to express. The provability of an explicit definition from a theory, under the conditions Tarski assumes, is necessary and sufficient for *definiens* and *definiendum* to require the same intuitive interpretation on pain of irrationality in the interpreter, and for this interpretation in turn to be settled by the interpretation of the primitive terms as far as possible. If the theory is in turn complete with respect to its specific terms, then definability amounts to complete capture of the intuitive meaning of an expression.

Tarski's treatment of definition concerns constraints on the association of intuitive meaning with the expressions of a deductive theory. Nothing in it, however, says anything about *which* intuitive concepts are associated with primitive expressions. For this Tarski needs a different sort of account: for a given intuitive concept, an account of the conditions under which a symbol that figures in certain theorems of a deductive theory expresses it. Here no general account of the structure of the theory is possible; what is needed is a treatment in terms of the analyses of these concepts themselves. We discussed this in the previous chapter with respect to Kotarbiński's account of connotation. The connotation of a term is the set of properties had by anything in its extension. To capture the connotation is for certain sentences to be assertible of anything to which the term is applied. If, for instance, a term F is to express the concept of being scarlet in a theory where G expresses the concept of being red, then $(\forall x)(Fx \rightarrow Gx)$ must be a theorem, since part of the connotation grasped by someone with the concept of a thing's being scarlet is that scarlet things are red.

The search for such an account of connotation suitable to the cases in which Tarski was interested—metamathematical concepts themselves—and its discovery in the case of "is true" and related terms, is the main philosophical achievement of Tarski's Intuitionistic Formalist conception of metamathematics, though we have already seen at least an assumed treatment of the analogous topic in his early articles on the consequence relation. Though in those articles Tarski has not yet faced the issue squarely, the treatment assumes a simple conceptual analysis of "is a consequence of" on which, e.g., consequence is reflexive and transitive. Beginning in the next chapter we will look at the issues as they arise with respect to semantic concepts, and in the subsequent chapter we will look at how Tarski arrived at an Intuitinistic Formalist treatment of the concept of truth that satisfied him.

Tarski's definitions themselves need not state conceptual analyses. Indeed, on Tarski's view definitions need not have anything to do with the analysis of a concept, and they need be neither intuitively analytic of it nor somehow intuitively true. What matters in a definition isn't its own relationship to the concept supposed to be expressed by its *definiendum*, but whether the deductive theory relative to which it is a definition comes, when it is added, to have theorems that either state the conception of the concept or otherwise constrain the defined term (ideally) to express just that concept. Definitions therefore need not have any particular epistemic or semantic features—they need not eliminate the defined concept in favor of others that are known with more certainly or are otherwise more basic.[22] For Tarski, a definition of a concept is always a sentence that, added to a theory, forces the defined term to express that concept.[23]

Consider here Tarski's definition of consequence for the language of the calculus of classes in [Tarski, 1983a]. Tarski first defines "consequence to the nth degree" in the obvious way: x is a consequence of the set of sentences x to the nth degree iff $n = 0$ and $x \in X$, or $n > 0$ and x is, e.g. the disjunction of two sentences that are consequences to the $n - 1^{th}$ degree, etc. (Definition 15 [Tarski, 1983a, 181]). He then defines consequence in the likewise obvious way: x is a consequence of X if it is a consequence to the nth degree for some n [Tarski, 1983a, 182]. The axioms of the early articles are then trivial consequences of the definition. Here it isn't the definition that is directly related to the conceptual analysis, but certain theorems; cf. Leśniewski's remarks discussed above (§ 1.2.2).

Both the axiomatic theory of consequence in [Tarski, 1983c] and the metatheory for the language of the calculus of classes supplemented by Tarski's definitions in CTFL suit "$\in Cn(X)$" to express the concept of being a consequence of the set of sentences X in virtue of the theories *as a whole* having the relevant theorems. The definition need not itself state the analysis, and it need not imply the analysis without help from the rest of the theory.

We now have some insight into Tarski's frequent use of enumerative definitions. It is often wondered how a "definition" that merely lists cases could possibly be suited for the analytical tasks that Tarski sets himself ([Field, 1972], [Davidson, 1990], [Mou, 2001] and many others). The answer is that the definition on its own isn't intended to carry out the analytical task; that job is done by the theory as a whole in implying the right theorems involving the *definiendum* once the definition is added. When Tarski is looking to "catch hold of the meaning of an old notion" [Tarski, 1986e, 665], it is always the deductive theory as a whole that does this, not a particular definition.

Another example from [Tarski, 1983a] makes the point well. Tarski's presentation of Definition 14 contains in microcosm all of the elements of Tarski's view about how an intuitive concept is captured in a deductive theory by adding a definition. He first writes:

> In the formulation of the definition of the concept of consequence I shall use, among others, the following expression: '*u is an expression obtained from the sentential function w by substituting the variable v_k for the variable v_l*'. The intuitive meaning (*inhaltliche Sinn, treść intuicyjna*) of this expression is clear and simple, but in spite of this the definition has a somewhat complicated form. [Tarski, 1983a, 180].

What follows is a fifteen-line definition that covers six cases. Clearly, a simple concept need not have a simple definition. So what suits it to capture the concept? Not any feature of the definition itself, but rather its consequences:

> For examle, it follows from this definition that the expresions $\iota_{1,1}$, $\bigcap_3(\iota_{3,1} + \iota_{1,3})$ and $\iota_{1,3} + \bigcap_2\iota_{2,3}$ are obtained from the functions $\iota_{2,2}$, $\bigcap_3(\iota_{3,2} + \iota_{2,3})$ and $\iota_{2,3} + \bigcap_2\iota_{2,3}$ respectively by substituting v_1 for v_2. But the expression $\bigcap_1\iota_{1,3}$ cannot be obtained in this way from the function $\bigcap_2\iota_{2,3}$ nor the expression $\bigcap_1\iota_{1,1}$ from the function $\bigcap_2\iota_{2,1}$ [Tarski, 1983a, 180].

The point here is that these consequences of the definition involve pairs of expressions that intuitively do or do not fall under the concept of an expression that results from another by substitution of a variable. We have here exactly the procedure we will later see treated at greater length with Convention T and a truth definition: Tarski announces the intention to capture a concept, introduces a rather complicated definition that doesn't look anything like a traditional conceptual analysis, and then judges the definition as worthy on grounds of the theory augmented by the definition having consequences that are intuitively true given the intended interpretation of the expression introduced and the assumed interpretation of the primitive terms of the theory.

3
Semantics

Tarski's views about meaning in the late 1920s and early 1930s were entirely in line with Intuitionistic Formalism as he conceived of it following Leśniewski and the closely related philosophy of language of Kotarbiński. The function of language is to convey the contents of thoughts and the meaning of a symbol is a content of a concept (in the case of "terms", expressions in the simple type-theoretic hierarchy) or its contribution to determining the meaning of such symbols (in the case of other "technical signs" [Tarski, 1983a, 167 note 1]). The primitive terms of a deductive theory get their meanings from the set of sentences held true, the theorems. Tarski's project was to craft deductive theories that expressed important metamathematical concepts via the constraints placed on their primitive terms by their theorems.

There is no hint anywhere in Tarski's early writings that denotation, satisfaction and truth play any role in the account of the significance of language, communication, or anything else of the sort; along Leśniewski's lines axioms and theorems, when taken to express thoughts, *seem* to be true, but for Leśniewski and Tarski this comes to nothing more than expressing beliefs that one has, given the conventions of the language. Ultimately, as noted in various places above, this emphasis on non-referential meaning-theoretic notions in the basic account of understanding and communication is part of Twardowski's legacy (e.g. [Twardowski, 1977, 8–9]).

Nevertheless one of Tarski's most celebrated contributions was to the development of the basic set of tools for thinking about meaning in terms of language–world relations, the concepts and techniques of referential, truth-conditional semantics. In the remainder of this work we will tell the story of Tarski's introduction of these tools and the development of his thought about them. Tarski initially didn't think of semantics as he introduced it as a contribution to the study of meaning at all. Tarski developed semantics not as a contribution to the basic theory of meaning, but as a bit of detail work in his project of giving Intuitionistic Formalist treatments of important metamathematical concepts. In this respect, the early development of semantics is simply a bit of work that extends his work on the consequence construed derivationally around 1930.

Unlike with his early work on consequence, however, Tarski's motivation for providing treatments of semantic concepts was in part to secure their legitimacy. Leśniewski and Carnap were openly hostile to semantic concepts, while their role in Kotarbiński's account is peripheral at most. Their understanding of variables, sentential functions and quantification was substitutional when it was clear. On the other hand, some of Tarski's own logical and mathematical work had focused in the late 1920s on what we can now recognize as early contributions to model theory by the American Postulate Theorists, and Löwenheim and Skolem, Hilbert and Ackermann, and others. Tarski's interest in giving definitions of semantic concepts was directed toward showing the legitimacy of these early model theoretic studies in the face of the skepticism of Leśniewski and Carnap by showing how terms expressing semantic concepts could be introduced into rigorous deductive theories by definitions and in accord with the constraints imposed by Intuitionistic Formalism.

3.1 Philosophical resistance

Thought about semantics before Tarski shows a strong bifurcation: on the one hand, there were developments in mathematical logic that involved the treatment of semantic concepts of truth and satisfaction as unproblematic, while on the other hand in philosophical circles there was a great deal of skepticism about semantic notions. We will discuss the latter here and the former in the next section. The blanket skepticism about semantic notions that one can find in the Vienna Circle, for instance, derived from the putatively "unscientific" character of the notion of truth that came from its frequent occurrence in metaphysical disputes (see [Soames, 1984, 415] and the references there) and from the idea that the conception of truth as correspondence committed one to some sort of incoherent conception of the confrontation of statements with facts (the primary motivation for eschewing semantic concepts ascribed, especially to Neurath, by Hempel [Hempel, 1935, 50–51]). In this section we will look in more detail at two topics directly connected with Tarski's interests: the interpretation of quantification and semantic paradox.

3.1.1   The quantifier

It is important to note how deeply mired in a substitutional conception of variables and quantification some of the figures to whom Tarski paid the most attention were.[1] One place we can see this is in Carnap's *Abriss der Logistik*, a work Tarski repeatedly cites:

> If a number of constituent signs are removed from a complex sign, there arise new complex signs by the insertion of different signs in the the positions thereby made free. To make the free positions themselves recognizable we could simply leave them empty, but we will as a rule place other signs in the free places that have only this purpose; these themselves have no meaning (*beteuten also selbst nichts*), but merely hold the positions open for the insertion of other signs. We say of the complete expression with these empty places that it signifies (*bezeichne*) a "function"; a variable that makes an open position recognizable are is called an "argument" (*Argument*) of the function. If constants are set in (or "substituted") for the variables, these constants are called "values" of the arguments, and the meaning (*Bedeutung*) of the resulting complex sign is called the "value" (*Wert*) of the function for the given argument values (*Argumentwert*)" [Carnap, 1929, 3] (my translation).

Here a function is an open sentence, the argument of a function is a variable, and the value of a variable is a constant. Some hope is given by the apparent distinction in the last sentence between the value of a function (= a closed sentence) and its *Bedeutung*, but this can hardly be taken seriously, since *Bedeutung* is, for Carnap, an "apparent" relation (*Scheinrelation*) [Carnap, 1929, 21]. So attention naturally turns to the apparently relational "*bezeichen*" in the passage. This, however, provides little comfort: if a function is something to which an open sentence *bears some relation*, then it doesn't make any sense to call the variables "arguments" of the function, rather the arguments should be that which is *bezeichnet* by the variables, but variables themselves, we have already been told, have no meanings at all.

Obviously what is needed here is the idea that the variables in an open sentence *range over* some objects. Without a semantic conception of variables and quantification, Carnap is left to flounder—though, being basically a clear thinker, the later construction of classes, cardinal numbers and so on is unaffected by confusion at the foundational level. As a further example, however, of the foundational unclarity, consider Carnap's definitions of the range of values of a function and of a set abstract:

"admissible argument values" are those which, when substituted into a complex expression, give it a meaning; for these the function has a value ... The range of admissible values (*Bereich der zulässigen Werte*) of a variable is called its "range of values" (*Wertbereich*) [Carnap, 1929, 3].

$\hat{z}(\varphi z)$ will mean: 'every value of z that satisfies (*befriedigen*) φ'; such an expression signifies a "class", in this case "the class determined by the sentential function φ" [Carnap, 1929, 16].

Taking this seriously along with the earlier definition of an argument value as a constant that can be substituted for a variable to produce a true, meaningful sentence, the range of values of a sentential function is a set of names, and the class determined by a sentential function is, again, a set of names. Classes, therefore, would seem to be sets of names. One might suspect that this is not what Carnap really wants to say but it is what he says. Indeed, and in hindsight astonishingly, when push came to shove Carnap actually took this seriously. The definition of analyticity for Language II of *Logical Syntax* goes to such length to make names do the work of objects in the account of quantification that Carnap would rather introduce nondenumerable sets of names to provide interpretations for higher order variables than simply to do things objectually (§6.2). We will return to this in chapter 6.[2]

Carnap is stuck with this, of course, because he holds that *Bedeutung* is a *Scheinrelation* "which does not appear at all in a purified language" [Carnap, 1929, 21] and, correlatively, that " 'true' and 'false' are undefinable concepts" [Carnap, 1929, 3]. The early pages of the *Abriss* cry out for Tarski's treatment: everything would go much more smoothly for Carnap if he had some way of talking about the relations between expressions and objects. However, without taking semantic relations seriously, one can hardly take variables and quantification as anything other than substitutional; the only notion of generality accessible to the semantic skeptic is substitutional generality.[3]

Kotarbiński likewise understands the variables in a sentential function substitutionally:

Sentential functions differ as to the kind of variables: there are term variables, sentential variables, conjunction variables, etc. ... a variable represents all constants of the appropriate category; hence a term variable represents all terms; a sentential variable, all sentences, etc. Since it has become the usage to call the constants represented by a variable the values of that variable (which has its source in the language of mathematics, where numerical variables are used, for example, in the case of the sentential function "$x = 2y$", where the "names of numbers" form the range of x and y), it can also be said that a variable represents all its values [Kotarbiński, 1966, 19].

The work does contain a rather extensive discussion of denotation, but treated only as an auxiliary feature of meaningful terms and entirely within the terms of traditional logic. The passage above makes clear that Kotarbiński grasped no possibility of a role for referential semantic notions in the interpretation of the basic expressive resources of logic itself.

Leśniewski's views on variables and quantification are a matter of dispute. His hostility to semantics (see below) makes clear that he couldn't have had anything to do with a properly objectual reading of the quantifiers nor with the corresponding conception of what a variable ranges over, but interpreters (e.g. [Küng, 1977]) reject Quine's suggestion [van Orman Quine, 1973, 99] that Leśniewski's understanding is substitutional. Some clue is given (a suggestion adopted by [Küng, 1977]) by his typical formulations in the early work of the form "for some meaning of the word 'a'" (e.g. [Leśniewski, 1992g, 125]), but when it comes the later work saying something responsibly would take us too far into the Leśniewski literature for our purposes, so I will leave the matter here.[4]

Tarski himself makes statements about variables and quantification, especially early on, that are substitutional or play with the idea that quantification somehow ranges over "meanings". For instance, in the early Leśniewskian "On the Primitive Term of Logistic" a reference to "quantifiers" is given this note:

> In the sense of Pierce ... who gives this name to the symbols '$\prod$' (universal quantifier) and '$\sum$' (particular or existential quantifier) representing abbreviations of the expressions: 'for every signification of the terms ... ' and 'for some signification of the terms ... ' [Tarski, 1983j, 1].

This makes fairly clear that at least in 1923 Tarski adhered to Leśniewski's view of quantification. Indeed a substitutional understanding of *sentential* variables persists in Tarski's writing all the way through the English version of the textbook, e.g. [Tarski, 1995, 38]. There is also a peculiar remark at one point in "The Concept of Truth" itself: "In the intuitive interpretation of the language, which I always have in mind here, the variables represent names of classes of individuals" [Tarski, 1983a, 169]. Intriguingly, though quantification in the object language gets an objectual treatment in the metalanguage, the "intuitive" reading of the object language is glossed substitutionally.

3.1.2 Paradox

Tarski attributes his concern with semantic paradox to Leśniewski [Tarski, 1983a, 154] and to nobody else. The lone attribution may well be apposite, as straightforward statements of the view that semantic concepts are not to be included in serious theory *because* of the possibility of paradox are in fact difficult to come across in the literature before Tarski. The liar paradox, of course, had been known since antiquity, but in the critical period following the discovery of Russell's paradox[5] the focus was on the set-theoretic paradoxes. With the advent of the Richard paradox [Richard, 1967] Peano initiated a tradition of distinguishing the semantic from the set theoretic paradoxes and writing the former off as "linguistic" or "epistemological" (e.g. [Fraenkel, 1928, 210]),[6] but I have been unable to find a clear statement of the view that semantics is to be eliminated *because* of the paradoxes in any of the places I would have expected to find it. For instance, although Tarski makes use of Łukasiewicz's formulation of the liar paradox [Tarski, 1983a, 157–8], Łukasiewicz himself does not seem to have taken it to cast doubt on the concept of truth, preferring instead a version of the familiar strategy of finding some reason that sentences that purportedly refer to or quantify over themselves are not well-formed (remark translated at [Woleński, 1994, 89]).

There is no other clue in Tarski's work as to who were the "specialists in the study of language" he had in mind among whom semantic concepts "for a long time ... have had an evil reputation" [Tarski, 1983a, 252]. The same issue arises with respect to "definable" in [Tarski, 1983f]: "mathematicians" are said to have an attitude of "distrust and reserve" with respect to the notion of definability, but as we will see actual mathematicians such as Skolem and the American postulate theorists displayed no such reticence. Since the opinion was quite common in Warsaw that a "mathematician" was anyone who studied axiomatized theories [Kotarbiński, 1966, 318ff] the reference could again be to Leśniewski and Tarski himself (cf. [Hodges, 2008, 128]). In 1935 Tarski again credits Leśniewski with appreciating the problem, but the context makes it relatively clear that what Leśniewski appreciated was that the semantic paradoxes show that "semantical concepts simply have no place in the language to which they relate" [Tarski, 1983b, 402], while the skeptical attitude long taken toward these concepts is attributed to no one in particular.

Tarski attributes his remarks on paradox in natural language to Leśniewski [Tarski, 1983a, 155] in lectures and discussions from 1919 onward. As [Betti, 2004] argues, some of the approach Tarski later takes up is foreshadowed in Leśniewski's 1913 paper "The Critique of the Logical Principle of the Excluded Middle", inasmuch as Lesniewski at least gives the T-sentences a role in the 'paradox of Epimenides' [Leśniewski, 1992c, 77] and holds that not all "'natural intuitions' of language" [Leśniewski, 1992c, 82] can be preserved in a "solution" to the paradox. However, the early Leśniewski provides a solution based on a ban on self-reference, applied in a context where sentence tokens are taken to be truth bearers [Leśniewski, 1992c, 80]. The "solution" therefore stands in the medieval tradition of Ockham and Buridan, and its closest contemporary relative is [Simmons, 1993], as Betti notes [Betti, 2004, 274]. The attitude that Leśniewski there takes toward the preservation of intuitions in the solution of the paradoxes also seems directly contrary to the one expressed at [Leśniewski, 1992h, 177–8], so it is left unclear exactly what sort of remarks Tarski heard in the years after 1919, though it seems fairly clear from the text that in 1913 Leśniewski hadn't yet arrived at the view attributed to him in ESS that "semantical concepts have no place in the language to which they are applied".

Aside from Leśniewski, the only figure making some sort of point to the relevant effect is Carnap, who in the *Abriss* appears to reject the inclusion of semantic terms in serious theory, if not on grounds of their giving rise to paradox, then at least taking the paradoxes to be a symptom of what is wrong with them. In a discussion of the branched theory of types from *PM* he writes:

> The branched theory of types was set forth because it was believed that certain antinomies of a special sort could not otherwise be avoided. (Here belong, for instance, the antimony of the smallest number that cannot be signified (*bezeichnet*) with one hundred characters, the "heterological" antinomy, and so on.) A closer examination appears to show that these antinomies are of, not a logical, but a linguistic sort, that is, they are only consequences of the deficiency of everyday language (*die Wortsprache*). (Namely, they all involve the apparent relation "means" ("*Bedeutens*"), which does not occur at all in a purified (*gereinigten*) language.) The problem of these antinomies is not yet solved; it hangs together with the problem of the thesis of extensionality (9e) [Carnap, 1929, 21] (my translation).[7]

The historical record, then, seems to indicate that skepticism about semantics based on the paradoxes came to Tarski primarily from Leśniewski with perhaps some help from Carnap; Tarski himself seems to be on the mark when he says later that "as far as I know" Leśniewski was the first to become fully aware of the fact that because of the paradoxes semantic concepts don't belong in the language to which they apply

[Tarski, 1983b, 402]. Aside from these points, the situation seems rather to have been that semantic concepts were ruled out on the grounds of a blanket consideration to the effect that they obviously expressed, as Carnap put it, *"Scheinrelationen"*; in this context the paradoxes were more a symptom than a source of the problem.

Leśniewski's influence is important, then, because only the worry about consistency motivates Tarski's project of definition. All of Tarski's strictly Intuitionistic Formalist goals could be met by axiomatizations. It was the problem posed by the paradoxes that led Tarski to the definitions for which he is now remembered.

3.2 Mathematical acceptance

The semantic perspective on meaning could be traced in Tarski's influences back through the algebraic logicians, in particular Pierce, whom Tarski singles out for mention in a discussion of the development of the appreciation of the role of variables in mathematics in the textbook [Tarski, 1995, 14]. More proximately, however, Tarski's influences in incorporating the basic notion of the satisfaction of a sentential function into metamathematics are the group known as the "American Postulate Theorists", and Löwenheim and Skolem.[8] As is well known [Vaught, 1986, 869–70] [Feferman and Feferman, 2004, 73] [Hodges, 2008, 120–4] in 1927–1929 Tarski conducted a seminar (actually, the exercise sessions attached to a seminar conducted by Lukasiewicz) in Warsaw that focused on Skolem's method of quantifier elimination and on related work by the American Postulate Theorist C. H. Langford. The point for us here is the issues that exercised Löwenheim and Skolem, as well as the postulate theorists, were ones that, against Carnap's scruples, demanded that semantic notions be somehow theoretically tractable, if not to anyone's knowledge definable in terms of something more uncontroversially acceptable by the standards of the time [Vaught, 1974, 161]. Löwenheim formulates both some of his basic definitions and the theorem that bears his name in terms of the satisfaction of expressions in a sense thoroughly familiar to us, though undefined and unremarked-upon [Löwenheim, 1967, 233–5]. Skolem, likewise, simply states the results in [Skolem, 1967] in terms of "domains", of which axioms "hold", or in which propositions are "satisfied". As with Löwenheim, the notions are used as though they were unremarkable.

Langford spends the early pages of [Langford, 1926] explaining issues of scope and multiple quantification, as well as the reduction of first-order formulas to normal form. The concern of the article is with how the truth-values of "first-order" functions are "determined" by sets of postulates (in the cases at issue, for linear order); the arguments go by way of establishing equivalences between first-order formulas of various forms and their normal forms. Like Löwenheim and Skolem, Langford shows no Carnapian or Leśniewskian scruple about the appearance of unreduced semantic primitives in the discussion. Langford speaks of "consequence", but in a way that we can recognize as semantic in accord with Tarski's later definition, explicitly distinguishing it from "strict consequence", which isn't remarked upon further:

> The general procedure of the proofs which follow may be outlined: Every multiply quantified function in n variables is shown to be equivalent, in view of (1)–(10) [Langford's postulates for dense order], to some function in two variables. This equivalence is, of course, material and not strict since it depends on (1)–(10). The function in two variables to which a function in n variables is shown to be materially equivalent is therefore in no sense another form of the same function. The functions have the same truth-value solely in view of (1)–(10) [Langford, 1926, 28].

Note the sensitivity to the semantic aspect of the issue: it's not that a function is equivalent to another in some sense of derivability or "saying the same thing", but simply that if the postulates (1)–(10) hold, it can be shown that they have the same truth value on all assignments to the variables.

Langford here stands in the tradition of the American Postulate Theorists, whose approach was thoroughly semantic or model-theoretic in the modern sense from the turn of the century on. Key postulate theorists such as Veblen (e.g. [Veblen, 1904, 344]) in turn cite the Peanists and Pasch.[9] Scanlan writes of their papers that they:

> contain numerous references to the European research literature of the period and are clearly meant to be contributions to lines of research that were ongoing in Europe. They nevertheless represent (1) clear formulations, both by explicit statement and by example, of a postulational point of view that is the author's own, and (2) advances, in some respects, over what was current in Europe [Scanlan, 1991, 982].

One can see this just after the turn of the 20th Century. [Huntington, 1902] reads in some ways as quite old-fashioned—e.g. at one place we get the explanation that an identity "$x = y$ indicates that the two symbols x and y are used to represent the same object" [Huntington, 1902, 266], the article is filled with talk of "rules of combination" for the objects that belong to "assemblages", and the notion of a "class" is given a gloss in terms of a condition—but the semantic perspective is fully in evidence; e.g. we have:

> Theorem II:—*Any two assemblages M and M' which satisfy the postulates 1–6 are equivalent; that is, they can be brought into one-to-one correspondence in such a way that a ∘ b will correspond with a' ∘ b' when ever a and b in M correspond with a' and b' in M', respectively* [Huntington, 1902, 277].

We saw above (§ 2.3) that Veblen's notion of categoricity is semantic. Veblen also explicitly distinguishes definition in the sense of a predicate's defining a set from definition in the sense of explicit definition:

> E. V. Huntington, in his article on the postulates of the real number system, expresses this conception by saying that his postulates are sufficient for the *complete definition* of essentially a single assemblage. It would probably be better to reserve the word *definition* for the substitution of one symbol for another, and to say that a system of axioms is categorical if it is sufficient for the complete *determination* of a class of objects or elements [Veblen, 1904, 346–7].[10]

Considerably later, both [Hilbert and Ackermann, 1928] and [Fraenkel, 1928], cited often by Tarski, work with the notions of satisfaction and a set of objects being a model of a set of sentences, treating them as unremarkable. Indeed, as will be at issue later (§ 7.3.2), Hilbert and Ackermann are able to pose the question solved by Gödel's Completeness theorem because they attend to the question of which sentences are true in models of all cardinalities [Hilbert and Ackermann, 1928, 68]. On the mathematical side, then, the picture is clear: mathematicians were working comfortably with the semantic notion of a system of objects satisfying some axioms in the decades before Tarski's work, and achieving fruitful results. There is no evidence that these figures harbored some misgivings about the semantic notions involved (cf. [Feferman, 2008, 80]).[11]

As a result of these trends, many authors at the time weren't pushing the substitutional accounts of quantification that Tarski's closest philosophical influences were. For example Hilbert and Ackermann, whom Tarski cites, have a passably clear semantic treatment:

> We still lack a symbolic expression for the generality of statements. To arrive at one, in the manner of mathematics we introduce alongside the signs for determinate objects (the proper names (*Eigennamen*)) also *variables x, y, z …* with which we can likewise fill the empty positions

in a function-sign. A determinate completion of an empty position is called a *value* of the variable in question.

The values of a variable are in general limited to determinate types of objects, which are themselves determined through the meaning (*Bedeutung*) of the function-sign. For example, the basic relation of plane geometry: "*der Punkt x liegt auf der Geraden y*" is represented by a function-sign with two arguments $L(x, y)$ Here only points are eligible as values for x and only lines as values of y [Hilbert and Ackermann, 1928, 45] (my translation).

The passage displays some peculiarities—e.g. there appears to be at least some use–mention confusion of Carnap's sort in the account of the value of a variable at the end of the first paragraph—but overall the account is the right one: the values of the variables in the second paragraph are points and lines, not their names. Thus ranging over objects and referring to them are important relations in the account. Such an explanation could not be offered by anyone who, like Carnap, thought there was something wrong with semantic concepts and who believed that the job could be done without them.

3.3   Intuitionistic Formalism in "On Definable Sets"

On the one side, we have Carnap, Leśniewski and others insisting that semantic notions simply can't play a role in serious theory; on the other, we have Skolem, the postulate theorists and others engaged in a good deal of serious theory that makes essential use of semantic notions [Vaught, 1974, 161]. It is sometimes wondered what set Tarski off on the project of defining truth [Feferman, 2008, 72–3], but surely the state of affairs occasioned just the reconciliation he attempted to effect. Compare here Gödel, who similarly was paying attention both to skeptics like Carnap and dogmatists like Skolem and who was moved to eliminate the appeal to intuitive semantic notions in [Gödel, 1967b] from the "rigorous" proof of his theorem. I thus think that Feferman understates the case at [Feferman, 2008, 72] when he holds that there was no "compelling *logical* reason" for Tarski's definitions and that Tarski had only "*psychological* and *programmatic* reasons" for the development of his truth definitions. Tarski's aim wasn't clarity for its own sake (contra [Feferman, 2008, 80]); his aim was to establish that semantic notions were immune from certain sorts of skeptical doubt. Importantly, note that Tarski never says that semantic concepts are unclear; on the contrary, he repeatedly emphasizes our "intuitive knowledge" of them [Tarski, 1983a, 154] and the fact that, of them, we can "give an account that is more or less precise in its intuitive content" [Tarski, 1983f, 111].

Intuitionistic Formalism meant that semantic concepts could be expressed effectively just in case their analyses could be expressed in the theorems of a deductive theory, while the worries about paradox required that this be done by definition rather than axiomatization in order to guarantee relative consistency. Thus Tarski had to tackle two problems at once: (a) the development of an Intuitionistic Formalist conception of the conditions under which a deductive theory had theorems that constrained one of its expressions to express some semantic concept, and (b) the development of techniques, acceptable in metamathematics as he construed it, for the introduction of such a term via a definition. We can recognize here already the two desiderata of stated in Convention T: the need for (a) is the need for a "materially adequate" treatment of semantic concepts, while the need for (b) is the need for "formally correct" definitions.[12]

However, Tarski was quicker to see the need (b) for explicit definition and to develop techniques for meeting it, than he was to work out (a) an Intuitionistic Formalist conception of what theorems needed to be provable in a deductive theory in order for that theory to have a term that expresses the concept of truth. Introducing an explicit definition of a term the intended interpretation of which is semantic into a non-semantic theory forces the intuitive interpretation of the term to be whatever is associated with the *definiens* by the definition. But it is one thing to see this, and another thing to have a view about what it takes for a term (primitive or defined) to have a formal role that suits it to express a semantic concept.

3.3.1 The intuitive notion of definability

The first semantic work in *Logic, Semantics, Metamathematics* is 1930's "On Definable Sets of Real Numbers". Here the question foregrounded seems to be in the spirit of the postulate theorists: given a theory and the Reals, which sets of Reals are defined by open sentences of the theory?[13] Importantly, some aspects of the truth-definition appear in the article, but without the fanfare accorded them in "The Concept of Truth in Formalized Languages". This is already significant: Tarski takes himself to be doing something important in the latter paper, but it cannot be merely setting out the truth-definition, since this appears in "On Definable Sets" with only brief commentary; indeed, he notes the equivalence of truth with satisfaction by an arbitrary sequence in a single sentence [Tarski, 1983f, 117].

"On Definable Sets of Real Numbers" is a more puzzling article than it might at first appear. Superficially, Tarski points out that semantic definition is the converse of satisfaction and introduces some of the mechanics of the treatment of the latter familiar to us from CTFL (though, as will be established below, far less than commentators often claim), and then proves some results about what sets of Reals are definable in particular theories—e.g., in Theorem 2, that first-order arithmetically definable sets of reals are finite sums of intervals with rational end points [Tarski, 1983f, 134]. The puzzle about the article concerns how Tarski could possibly think that doing what he does in the article amounts to doing what he says at the outset he is going to do. What he *says* he will do is this:

> The distrust of mathematicians toward the notion in question [definability] is reinforced by the current opinion that this notion is outside the proper limits of mathematics altogether. The problems of making its meaning more precise, of removing the confusions and misunderstandings connected with it, and of establishing its fundamental properties belong to another branch of science—metamathematics.
>
> In this article I shall try to convince the reader that the opinion just mentioned is not altogether correct. Without doubt the notion of definability as usually conceived is of a metamathematical origin. I believe that I have found a general method which allows us to construct a rigorous metamathematical definition of this notion. Moreover, by analyzing the definition thus obtained it proves to be possible (with some reservations to be discussed at the end of § 1) to replace it by a definition formulated exclusively in mathematical terms. Under this new definition the notion of definability does not differ from other mathematical notions [Tarski, 1983f, 110–111].

The puzzle is that though Tarski certainly sets up a correlation between his theory of first-order arithmetic and the set theoretic construction with which he pairs it, there does not at first reading appear to be anything in the article that would justify the claim that he has captured the "meaning" ("*sens*" [Tarski, 1986g, 519]) of "definable set of real numbers"—for, surely, that meaning is "metamathematical", while Tarski's "mathematical" definition of a certain set of sets of Reals as the closure of certain others under certain operations precisely is not. Furthermore, as we will see shortly, crucial aspects of the semantic treatment in CTFL are simply missing from the article, thereby making it questionable whether in the theory itself—as opposed to the mind of the mathematician contemplating it—anything semantic is expressed at all.

The determination to introuduce a term that expresses a semantic concept into a deductive theory in Intuitionistic Formalist fashion can be seen in one of the most philosophical passages in Tarski's corpus:

> The problem set in this article belongs in principle to the type of problems which frequently occur in the course of mathematical investigations. Our interest is directed towards a term of which we can give an account that is more or less precise in its intuitive content (*contenu intuitif*), but the meaning of which has not at present been rigorously established, at least in mathematics. We then seek to construct a definition of this term which, while satisfying the requirements of methodological rigor, will also render (*établie*) adequately and precisely the actual meaning of the term [Tarski, 1983f, 111–112] [Tarski, 1986g, 521].

Note several things about this passage so far. First, we have here a philosophical take on mathematics and mathematicians being concerned with something like conceptual analysis: clarifying an "intuitive content" is a *mathematical* goal in this passage. Since we know from Kotarbiński that anyone who studied deductive theories was a "mathematician", and since we know from Leśniewski that one of the main goals of a deductive theory is to allow the expression of thoughts involving the concepts expressed by its primitives, the passage can be taken to mean that there has been doubt as to whether semantic notions could be expressed in a deductive theory in Intuitionistic Formalist fashion. Tarski surely has Leśniewski in mind here. Second, note the clear appearance of the Intuitionistic Formalist goal of expressing an intuitive concept in a deductive theory in the passage: there is a concept we intend to express by a term; we have a good idea of what it is, yet this somehow compatibly with a lack of "rigor". We have the notion of "adequately and precisely" (*avec justesse et précision* [Tarski, 1986g, 521]) capturing "intuitive content", and also an allied requirement that we do so "while satisfying the requirements of methodological rigor"—precursors of Convention T's "materially adequate and formally correct". The gap between our intuitive notion and this sought-for rigor is to be closed by constructing a formal definition. So far this is all orthodox Intuitionistic Formalism. Let us continue with the passage:

> It was just such problems that the geometers solved when they established the meaning of the terms 'movement', 'line', 'surface', or 'dimension' for the first time [Tarski, 1983f, 112].

The late 19th and early 20th century geometers Tarski is thinking of were those who gave rigorous axiomatic treatments. As we know from our discussion of Intuitionistic Formalism, capturing "intuitive content" in the way that concerns Tarski is a matter of setting up a deductive theory that implies theorems involving a symbol that constrain it to express this content. Continuing:

> Here I present an analogous problem concerning the term 'definable set of real numbers' [Tarski, 1983f, 122].

Given all this, the task set for the article is the expressive one Tarski has been interested in for metamathematical concepts, and formal definition *within* a deductive theory will be the focus of attention. Although the article is about semantic definition, its stated goal is an explicit formal definition that captures a certain concept. We can see, here as elsewhere, the extent to which Tarski's main interest in the early 1930s is in this expressive task, with the semantic content of the article, such as it is, only coming in in service of it. Continuing again:

> Strictly speaking this analogy should not be carried too far. In geometry it was a question of making precise the spatial intuitions acquired empirically in everyday life, intuitions which are vague and confused by their very nature. Here we have to deal with intuitions more clear and conscious, those of a logical nature relating to another domain of science, metamathematics [Tarski, 1983f, 122].

The announced task is to analyze a *concept* from metamathematics—not merely to *do* some metamathematics by setting up the formal semantics. Rather, the formal semantics serves the conceptually analytic goals. Continuing:

> To the geometers the necessity presented itself of choosing one of several incompatible meanings, but here arbitrariness in establishing the content of the term in question is reduced almost to zero [Tarski, 1983f, 122].

The reference here is presumably to axiomatic studies of geometry such as those of Hilbert [Hilbert, 1971] and Veblen [Veblen, 1904]. I take the point of the passage to be that in geometry intuitive concepts of space acquired in everyday life don't select among various geometries and that hence the conceptual analysis of geometric notions is controversial. By contrast there is no similar dispute about "definable" and hence we have good hope for an uncontroversial conceptual analysis, leaving us only the task of crafting a deductive theory that constrains one of its primitive terms to express the concept so analyzed.

> I shall begin then by presenting to the reader the content of this term, especially as it is now understood in metamathematics. The remarks I am about to make are not at all necessary for the considerations that will follow—any more than empirical knowledge of lines and surfaces is necessary for a mathematical theory of geometry [Tarski, 1983f, 122].

Since the topic discussion of which is "not at all necessary" here is the same one that is "of capital importance" at [Tarski, 1983f, 129], the point of this passage must be the standard Intuitionistic Formalist one that a deductive theory can be understood by someone who doesn't know what concepts its primitive terms are intended to express. This is clear in the next sentence:

> These remarks will allow us to grasp more easily the constructions explained in the following section and, above all, to judge whether or not they convey the actual meaning of the term.

The completed theory will supposedly express as well as possible the "intuitive content" of "definable set of real numbers"; the point here is that it does so whether or not one *knows* that it does, but that of course one cannot know that it does if one doesn't have this intuitive content in mind, and so the reader will be reminded of this content before the construction begins.

We can pause here to note an issue that we discussed in Chapter 2: the relativization of Tarski's definitions to a particular object language. As with the notions of sentence and consequence, Tarski's treatment of definability is relative to a particular deductive theory of arithmetic, but he systematically suppresses this relativization and is happy to speak of defining "the" concept of definability, rather than of defining the concept of definability in that formulation of arithmetic. The points made there carry over here: on Tarski's conception of a concept linguistic concepts must be relativized to a language to have conditions of applicability and thus extensions. There is only one concept of a definable set of real numbers, but it is applied only relative to an object language. Tarski pushes this relativization into the commentary on his metatheory because the notion of being in a language isn't metamathematically tractable.

3.3.2   Defining definable sets vs defining "Defines"

Readers of CTFL will expect certain things from here forward in ODS that the article simply doesn't supply. In particular, one would expect

1. A rigorous distinction between object language, metalanguage, and metametalanguage
2. A criterion of adequacy for a successful definition, formulated in the metalanguage, 'x is a definable set of real numbers' for the object language
3. Meticulousness about use and mention; in particular, a clear distinction between expressions of the metalanguage that refer to or describe object language sentences, and these sentences themselves.
4. A conception of definitions of semantic terms on which they are in some sense "logical sums" of "partial definitions" that trade between use and mention [Tarski, 1983a, 264].

ODS, however, contains none of these things. The only nod to them is this passage:

> Let us now consider the situation metamathematically. For each deductive system it is possible to construct a particular science, namely the 'metasystem', in which the given system is subject to investigation. Hence, to the domain of the metasystem belong all such terms as 'variable of the *n*th order', 'sentential function', 'free variable of a sentential function', 'sentence', etc. On the other hand nothing forbids our introducing into the metasystem arithmetical notions, in particular real numbers, sets of real numbers (*des notions d'Arithmetique, donc en particulier celle de nobre réel, d'ensemble de nombres réels*), etc. By operating with these two categories of terms (and with general logical terms) we can try to define (*préciser*) the sense of the following phrase: '*A finite sequence of objects satisfies a given sentential function*' [Tarski, 1983f, 116] [Tarski, 1986g, 523].

Note that "define" for "*préciser*" obscures the important Intuitionistic Formalist distinction between "making precise" an intuitive meaning and introducing a definition that augments a deductive theory to contain a term that has that meaning so precisified and that Woodger has dropped the interest in introducing the "*notions*" of real numbers and sets of them.

The construction in § 2 goes, in some respects, in a now-familiar way. First, Tarski augments simple type theory with "the specific terms of the arithmetic of the real numbers", sentential functions v, μ and

σ, receiving the interpretations "$x_{/}^{(k)} = 1$", "$x_{/}^{(k)} \leqslant x_{/}^{(l)}$" and "$x_{/}^{(k)} = x_{/}^{(l)} + x_{/}^{(m)}$", the subscripts expressing order and the superscripts enumerating the variables within an order [Tarski, 1983f, 114]. Tarski then points out that each of the primitive sentential functions determines a set of sequences of objects that satisfy it and he uses the set abstraction operator to form expressions for these sets; we can represent these here, ignoring the point that really sets of ordered pairs are determined as Tarski does things for one-place sentential functions ([Tarski, 1983f, 117]; the sequences are just book-keeping for the enumerated variables), as $\{x \mid x = 1\}$, $\{(x,y) \mid x \leqslant y\}$ and $\{(x,y,z) \mid x = y + z\}$. Correspoding to the classical logical "fundamental operations" are operations on these sets of the familiar sorts—union for disjunction and so on. The first-order definable sets Reals in STT augmented by the primitive sentential functions are then simply the closure of the primitive sets under the operations (Def. 9, 128) and the sets of Reals definable are those to be found in any one-term sequence (Def. 10, 128).[14]

It is sometimes claimed (e.g. [Hodges, 2008, 125]) that the article contains the first compositional semantics for a formal language.[15] Though it is true that some of the technical apparatus for this is in place in the article it is less clear whether what happens in the theory discussed in the article really amounts to as much. For rather than introducing some systematic means of *reference to* sentences of his deductive arithmetic theory in his metatheory, along with translations of its sentences into the language of the set-theoretic construction and then explicitly defining "x is the set of real numbers defined by y" in this metatheory, Tarski simply introduces set-theoretic vocabulary into what appears to be the arithmetic theory itself, defines certain "primitive sets" and "operations" on them, and then defines a set $\mathcal{D}$ of sets of real numbers as the closure of the primitive sets under the operations [Tarski, 1983f, 128] (ignoring some details). When we turn to the question of whether the sense of "definable" has been captured in the extended theory, we get this announcement:

> Now the question arises whether the *definitions just constructed* (the formal rigour of which raises no objection) *are also adequate materially* (juste au point de vue matériel); in other words, *do they in fact grasp* (saisit) *the current meaning of the notion as it is known intuitively* (le sens courant et intuitivement connu de la notion)? Properly understood, this question contains no problem of a purely mathematical nature, but it is nevertheless of capital importance for our considerations [Tarski, 1983f, 129] [Tarski, 1986g, 538].

Anticipating the structure found in CTFL, we would expect at this point some argument to the effect that "$\in \mathcal{D}$" has been introduced in some way that implies clearly semantic "partial definitions" in some plausibly exhaustive way; here the partial definitions would be of the form:

x is defined by $\ulcorner \varphi \urcorner$ iff $\forall y(y \in x \Longleftrightarrow \varphi(y))$

This, however, is not what we get. Tarski first states the following:

> We have in fact already established that there is a strict correspon-
> dence between primitive sets of sequences and the fundamental
> operations on these sets on the one hand, and the primitive sentential
> functions of order 1 and the fundamental operations on these expres-
> sions on the other ... From these facts, Def. 9, and the definitions of
> § 1 of a sentential function and its order, we can show without diffi-
> culy that the family $\mathcal{D}f$ is exactly the family of sets of sequences which
> are determined by sentential functions of order 1. It follows almost
> immediately that the family $\mathcal{D}$ coincides with that of the definable
> sets of order 1 in the sense of § 1 [Tarski, 1983f, 129].

What this says is that we can quite easily prove that the closure of the primitive sets under the fundamental operations is the set of sets defined by sentences open on a single first-order variable. Nothing, however, in the "metamathematical" § 1 of the article shows us how to *eliminate* "definable" in terms of something non-semantic. The described proof, in turn, simply shows us that the (first-order) definable sets metamath-ematically construed are exactly the sets in the closure of the primitive sets under the fundamental operations of § 2. That shows that the "math-ematical" Definition 10 gets the extension right. It does not, however, show what Tarski claims in the article will be shown, namely, that the *concept* of definability will be expressed in a theory with mathematical primitives. The sketched proof thus does not, so far, address the issue of "capital importance". To do that would require some argument that "$\in \mathcal{D}$" is caught up in theorems of the introduced deductive theory that constrain it to express the intuitive concept of definability. A mere proof of extensional adequacy doesn't do this job.[16]
Tarski therefore continues:

> If we wish to convince ourselves of the material adequacy (*justesse
> matérielle* [Tarski, 1986g, 538]) of Def. 10 and of its conformity with
> intuition without going beyond the domain of strictly mathematical
> considerations, we must have recourse to the empirical method.[17] In
> fact, by examining various special sets which have been arithmeti-
> cally defined (in the intuitive sense of this term), we can show that

all of them belong to the family D; conversely, for every particular set belonging to this family we are able to construct an elementary definition. Moreover, we easily notice that the same completely mechanical method of reasoning can be applied in all the cases concerned.

The following shows what this method involves: Let A be any set of numbers elementarily defined with the help of the primitive notions '1', '$\leqslant$', and '+'. The definition of the set A may be put in the form:

$$A = E_{x^{(k)}} \phi(x^{(k)})$$

The symbol '$\phi(x^{(k)})$' here represents a certain sentential function containing the variable '$x^{(k)}$' of order 1 as its sole free variable. It may also contain a series of bound variables '$x^{(l)}$', '$x^{(m)}$', etc., provided that all these variables are also of order 1 (since if this were not the case the definition could not be called elementary). As we know from logic, the function '$\phi(x^{(k)})$' can be constructed in such a way that it contains no logical constants other than the signs of negation, of logical sum and product as well as the universal and existential quantifiers. In the same way we can eliminate all the arithmetical constants except the signs 'v', 'μ', and 'σ', corresponding to the three primitive notions.

It is not difficult to show in each particular case that the formula [displayed above] can be transformed in the following way. The sentential functions '$v(x^{(k)})$', '$\mu(x^{(k)}, x^{(l)})$', and '$\sigma(x^{(k)}, x^{(l)}, x^{(m)})$' are replaced by symbols of the form 'U_k', '$M_{k,l}$', and '$S_{k,l,m}$', respectively; the signs of logical operations by those of the corresponding operations on sets of sequences, introduced in Defs. 5–7; and finally the symbol '$E_{x^{(k)}}$' by 'D'. By this transformation the formula [above] takes the form

$$A = D(S)$$

where in place of 'S' there is a composite symbol, the structure of which shows at once that it denotes a set of sequences of the family Df with counter domain consisting of a single element. By applying Def. 10 we conclude at once that the set A belongs to the family D.

I shall now give some concrete examples:

1. Let A be a set consisting of the single number 0. We see at once that $A = E_x(x = x + x)$, i.e. that $A = E_x\sigma(x, x, x)$. By transforming this formula in the way just described we obtain $A = D(S_{1,1,1})$, so that $A \in D$. [Tarski, 1983f, 129–131].

Tarski adds several more examples and then finishes with a flourish: "As a consequence of these considerations, *the intuitive adequacy of Def. 10 seems to be indisputable*" [Tarski, 1983f, 132].

How does this help to answer the question of "capital importance" of whether "$\in \mathcal{D}$" expresses, within the deductive metatheory, the concept semantic definition, thereby making it "indisputable" that Def. 10 is intuitively adequate? Note that the identities proven here by the "completely mechanical method" of replacing arithmetic vocabulary with appropriately selected set-theoretic vocabulary and the set-abstraction operator with "$\mathcal{D}$" involve no interplay between use and mention. They are all of the form

$$\{x \mid \varphi(x)\} = \{x \mid \psi(x)\}$$

where φ is an arithmetic predicate and ψ is a set-theoretic one. The identities all establish that a set of real numbers specified by abstraction from an arithmetic condition is the range (officially, in the article, the "domain" [Tarski, 1983f, 121 note †]) of a set of one-termed sequences specified in terms of Tarski's primitive sets and fundamental operations.

Part of the problem here is that Tarski seems to have bamboozled himself with his own metalinguistic explanation of his set-abstraction operator which, he claims, when prefixed to a sentence ϕ open on one first-order variable, "denotes the set of all objects that satisfy the condition ϕ" [Tarski, 1983f, 121]. (This explanation corresponds to Carnap's at [Carnap, 1929, 16].) The operator, however, doesn't actually do that: it simply forms a referring expression from an open sentence. What happens in the article, however, is that for the philosophical purposes of "convincing ourselves of the material adequacy of Definition 10" the set abstraction operator is viewed as introducing a quotational context—this despite Tarski's qualms about quotation [Tarski, 1983a, 159ff]. Viewed this way, the identities provable by Tarski's "completely mechanical method" do have metalinguistic content; rather than the above, they are of the form:

$$E(\ulcorner \varphi(x) \urcorner) = \{x \mid \psi(x)\}$$

What is really going on here is actually something somewhat related to the transformation to which Tarski subject's Padoa's method in [Tarski, 1983k]. Just as before, in Tarski's "justification" of the method by formulating Padoa's condition as a deductive one, the crucial metamathematical content is in the mind of the mathematician who contemplates a deductive theory while "varying the meaning" of the primitive terms—in ODS too the semantics is still happening in the mind of the mathematician: contemplating a deductive theory that contains both arithmetic and set-theoretic vocabulary, one can *view* the set-abstraction operator as introducing a quotational context and thereby take the identities proven by the "mechanical method" as having metalinguistic content. But within the contemplated theory itself, we simply have mathematical equivalences—indeed, this is assumed by the "mechanical method", which would otherwise involve obviously invalid substitutions of expressions within quotational contexts. It is this double interpretation of the set abstraction operator that underlies the rather surprising lack of fastidiousness about issues of object language, metalanguage and metametalanguage, as well as concomitant issues about use and mention, in § 2 of ODS. In particular, the reader will note that while CTFL carefully distinguishes, within the metalanguage, between translations of object-language expressions and structural descriptive terms that refer to them, this apparatus is entirely missing in ODS, since its job has been taken up by use–mention ambiguity in the interpretation of the context introduced by the set-abstraction operator.

Nevertheless, an early, inchoate grasp on Convention T seems to be at work in Tarski's "completely mechanical method", since the idea is that once we take a one-place first-order arithmetic open sentence in normal form, its specifically arithmetic vocabulary can be systematically replaced by related set-theoretic vocabulary: the primitive sets for the primitive sentential functions and the relevant set-theoretic operations for the logical operations on sentential functions. The thought here is clearly the correct one that will later inform Convention T: "$\{1\}$" translates "$v(x^{(k)})$", "$\overset{\circ}{+}$" occurring between two expressions that denote sets translates "or" as it occurs between two one-place open sentences, and so on. However, Tarski hasn't yet fully grasped the later idea that the essence of capturing a semantic concept is a certain interplay between use and mention. Instead, what we get here is an invitation to think about a deductive theory in which certain identities can be proved, and to view the relevant identities as having metalinguistic content. The advance of CTFL will be to bring this interplay into the theory itself, just as Tarski's treatment of Padoa's method brings the "variation in meaning" it involves out of the mind and into the formalized aspects of the deductive theory itself.

Indeed, since genuine metamathematical content is really missing from the section, what we get is not a definition of a term that expresses a *relation* of definability between a set of reals and an arithmetic theory, but rather a set of reals that happens to be the set of reals definable within a particular arithmetic theory.[18] ODS therefore simply does not succeed in capturing the intuitive notion of definability adumbrated in § 1 within the mathematical theory of § 2. The relevant topic, capturing concepts of word–world relations, simply hasn't come clearly into view yet. Tarski is groping toward the idea that theorems with the now-familiar interplay of use and mention (or use, translation and mention) are required, but the role of the appeal to translation and the crucial place of fastidiousness about use and mention, and hence object-, meta- and metametalanguage, isn't yet in properly in focus. This isn't to say that Tarski wasn't elsewhere careful about object- and metalanguage issues, or use and mention before 1929. What he didn't clearly conceive at the time of composing ODS was that properly capturing a semantic concept applied to some object language within a deductive metatheory was going to require him to incorporate the apparatus needed for such clarity into the metatheory itself, and thus more clearly to distinguish the metametalanguage from the metalanguage. A few years later Tarski puts the thought as something quite clear and obvious:

> For each of these [semantic] concepts we formulate a system of statements, which are expressed in the form of equivalences and have the character of partial definitions; as regards their contents, these statements determine the sense of the concept concerned with respect to all concrete, structurally described expressions of the language being investigated. We then agree to regard a way of using (or a definition of) the semantical concept in question as materially adequate if it enables us to prove in the metalanguage all the partial definitions just mentioned. By way of illustration we give here such a partial definition of the concept of satisfaction:
>
> *John and Peter satisfy the sentential function 'X and Y are brothers' if and only if John and Peter are brothers*
>
> It should also be noted that, strictly speaking, the described conventions (regarding the material adequacy of the usage of semantical concepts) are formulated in the metametalanguage and not in the metalanguage itself [Tarski, 1983b, 404–5].

These thoughts not yet in view, ODS breaks down at just the point where CTFL succeeds. Tarski hasn't yet done for semantics what he did for Padoa's method by 1926: bring the relevant concepts into the deductive theory in the form of a condition on what must be derivable for its expressions to express the target metamathematical concepts.

The work in ODS dates to sometime in 1929 [Hodges, 2008, 124]. The French version of the article appeared in 1931, but Kuratowski reported on its contents in mid-1930. So what we can conclude is this. By the spring of 1930, Tarski understood his task as being to construct definitions that introduced terms that expressed semantic concepts into a mathematical theory for a fixed object language and he had some of the tools required for this in hand. Less clear before mid-1930 was a specification of the conditions under which this had been accomplished. It is clear from the opening pages of ODS that he had some idea that he wanted to capture the "intuitive meaning" of semantic notions in mathematical theories, but it is also clear from the treatment of this matter of "capital importance" in the article that at the time of writing he hadn't fully specified either what this task amounted to, or how he was to show that it had been accomplished. Since the dating places "On Definable Sets" right around the period when Kotarbiński's *Elementy* appeared, we can say that Tarski had an unclear conception of a philosophical project that was waiting to be developed. With the treatment of truth in the *Elementy* Tarski was able to move forward [Hodges, 2008, 129].

Nevertheless, the project embarked upon with ODS is clear: craft a deductive theory with a structure such that certain primitive or defined expressions express semantic concepts in Intuitionistic Formalist fashion. The issue of "capital importance" is just this, and Tarski tries his best to show that his definition of "definable set" is in that sense "intuitively adequate". In late 1929, when Tarski was beginning to develop the formal techniques for which he is now famous, their purpose was to contribute to the expression of semantic concepts by the expressions of a deductive theory. Semantic expressions in ODS are precisely not part of the implicit "theory of meaning" of the article; that theory of meaning is the same one Tarski adhered to throughout the 1920s and up until 1934, the Intuitionistic Formalist one that stressed the expression of intuitively valid thoughts and the contents of concepts in deductive theories. We can also note for later use that in ODS the notion of consequence remains the proof-theoretic one specified by intuitively valid inferences [Tarski, 1983f, 116]. Semantics in 1930 for Tarski had nothing to do with the analysis of meaning or the relation of logical consequence.

4
Truth

Tarski's interest in the early 1930s is in the expression of metamathematical concepts within mathematical deductive theories. Having expressed the concept of logical consequence as he understood it at the time, Tarski moved on to semantic concepts, motivated by the clash between philosophical scruples and mathematical practice with which we opened the previous chapter. Since Intuitionistic Formalism requires that an expression that expresses a semantic concept be bound up in theorems that constrain it to express that concept, the exercise requires a conception of what these theorems are. This requirement appears in an inchoate way in early 1930 as the requirement that these eliminative definitions be "intuitively" or "materially" adequate, but in late 1929 or early 1930 Tarski lacks any particularly clear idea of what this amounts to in the case of semantic concepts, though he shows some signs of groping toward the right idea in his treatment of the "completely mechanical method" of translating arithmetic formulae into set-theoretic ones—a clear precursor of CTFL's appeal to translation.

We thus have two primary expectations for "The Concept of Truth in Formalized Languages". First, Tarski, as is shown by the continuation of this interest as late as 1934's "Definability of Concepts", will in the article be interested in the first instance in the expression of intuitive semantic concepts in a deductive theory. Second, the *magnum opus* should extend and correct the work of "On Definable Sets" by developing a more articulate conception of what it takes to express intuitive semantic concepts within a deductive theory, whether the expressions in question are defined or not. The two most celebrated aspects of the article fulfill these twin expectations: Convention T is the clear conception that replaces the fumbling treatment of intuitive adequacy in "On Definable Sets", while the method of providing a definition that is adequate according to

Convention T provides defined terms that have the sought-after features. Extensional semantics, on the other hand—strangely enough, given the reception of Tarski's work by posterity—will remain in the back seat throughout the article, its only interest being its role in the explicit definition of semantic terms in an intuitively adequate way.

The first three paragraphs of the work, puzzling without an understanding of Tarski's Intuitionistic Formalism, thus read exactly as we should now expect: the goal is to show how to craft a deductive theory containing a certain term so that that term is constrained to express a familiar concept. On the usual reading of Tarski, on which a definition is supposed *itself* to express a "thorough analysis" of a concept, it is a mystery why Tarski would both set out to define truth and deny that he intends a thorough analysis thereof. The passage, however, institutes just the structure we have seen in the early pages of "On Definable Sets". A formal theory, understandable in its own terms, is to be given a certain structure. Metatheoretically, this structure can be seen to suit one of its signs for the expression of an intuitive concept. Some remarks will be made on this intuitive concept in order to "remind the reader" of its basic features. However, that the formal theory has the requisite structure is a matter of its having certain sentences as theorems; if the theory achieves this definitionally, this doesn't mean that the definition in question itself states an analysis of the target concept.

4.1 Convention T

4.1.1 Terminological notes

The announced aim of "The Concept of Truth in Formalized Languages" is, in Woodger's translation, the development of a "materially adequate and formally correct definition of the term 'true sentence'" [Tarski, 1983a, 152]. "Materially adequate" here is Woodger's rendition of Blaustein's "*sachlich zutreffende*", which in turn renders the Polish "*merytorycznie trafnką*". This introduces an important issue in the translations. "*Merytoryczny*" is cognate with "merit". Rudolf Haller renders Tarski's "*merytoryczny*" in German as "*meritorisch*" [Tarski, 1992, 3], making clear the etymological connection with "meritorious", but Blaustein renders "*merytoryczny*" as "*sachlich*". Contrast here Tarski's own French of "On Definable Sets", in which he is perfectly happy to use "*intuitif*" (e.g. [Tarski, 1986g, 521, 538]) repeatedly to express the same idea: capturing the ordinary meaning of a term. Following Tarski's French, the German "*intuitiv*" and thence the English "intuitive" would have been perfectly acceptable.[1]

Woodger then brings Blaustein's "*sachlich*" into English as "materially". This transformation then lays the foundation for a longstanding confusion about Tarski's aims, and in particular about the purpose of Convention T: the confusion according to which the point of Convention T is to ensure that the defined truth-predicate is *extensionally* adequate, since "material" suggests "material conditional" which in turn brings to mind conceptions of extensional logic. One of my primary aims here will be to argue that this is not right. Convention T contributes to Tarski's Intuitionistic Formalist project: it states a condition under which an intuitive concept has been properly expressed in a deductive theory. That said, Woodger has his reasons: [Tarski, 1983f] treats "*intuitif*" and "*matériel*" as synonymous in several passages, e.g. [Tarski, 1986g, 538] where the question of whether Def. 10 possesses "*justesse matérielle*", answered affirmatively, gives rise to the claim that the definition has "*justesse intuitive*". Moreover, Woodger, working in the 1950s, had [Tarski, 1986e] at his disposal, and Tarski's own English uses "material" for the notion, so "material" it was for the 1933 work as well. Nevertheless, the important notion is being intuitively adequate (recall here also that additional appeals to intuition in Tarski's Polish are hidden by the German and English translations' substituion of linguistic usage for intuition (§ 1.4.3)). We will do best to keep this in mind by sticking to "intuitive" for the notion.

This brings us to a second issue with Blaustein's translation and Woodger's translation of it [Hodges, 2008]. A definition of truth should be intuitively adequate and formally correct. There are thus two evaluative notions at work: being logically in order (*poprawny*), and capturing the concept (*trafny*). Unfortunately, Blaustein varies the translation of the latter, and Woodger then gets confused about which two of Blaustein's three terms go together:

> Blaustein translates *poprawny* as *korrekt*, while he translates *trafny* sometimes as *zutreffend* and sometimes as *richtig*. Woodger realizes that there are only two concepts involved, but he mistakenly thinks that *richtig* is a variant of *korrekt* and not of *zutreffend*. So he translates both *richtig* and *korrekt* as 'correct'; in one place where Blaustein has *korrekte und richtige Definition*, Woodger just writes 'correct definition' [Hodges, 2008, 117].

Hodges gives a full list of the places where Woodger's translation needs to be corrected; I'll note these places below where they are relevant to our concerns. The problem is quite serious: in nearly every one of the central discussions of § 3 and § 4 where intuitive adequacy is at issue,

Woodger has "correct" where Tarski's Polish has "*trafny*" (pages 214 and 224 are prime examples). This greatly obscures the extent to which Tarski repeatedly returns to the theme of capturing the intuitive concept of truth, leaving him looking much more concerned with matters that are somehow "formal" than he actually is since the reader is primed to hear "correct" as "formally correct". That said, Blaustein seems to have an excuse, having simply adopted Tarski's own indifference between "*richtig*" and "*zutreffend*" as German translations for "*trafny*": In the Postscript to CTFL Tarski has "*sachlich zutreffend*" at [Tarski, 1986a, 192] and "*richtig*" at [Tarski, 1986a, 191].[2] Likewise [Tarski, 1986c, 262, 264] has "*sachlich zutreffend*", but Tarski switches to "*sachlich richtig*" at [Tarski, 1986c, 263].

4.1.2   Truth in the Lvov–Warsaw school

In an encyclopedic treatment at this point I would discuss analyses of truth before Tarski's time in order to set the analysis of the concept of truth that Tarski takes as the target for expression in a deductive theory in full historical context. This is not possible here at acceptable length and so I will turn directly to the tradition that leads to Convention T.[3]

Two ideas were present in the Polish tradition in Warsaw during Tarski's time: endorsement of certain formulations taken to be formulations of the correspondence theory of truth, and the view that the concept of truth is somehow captured by the T-sentences. Endorsement of the correspondence theory, its association with 1011b1 of Aristotle's *Metaphysics*, and criticism of conceptions of correspondence in terms of facts and propositions—familiar features, all, of Tarski's discussions ([Tarski, 1983a, 155], [Tarski, 1983b, 404])—stem from Twardwoski, quoted at [Wolenski et al., 2008, 21–43 23]:

> An affirmative judgment is true if its object exists, a negative judgment, if its object does not exist. An affirmative judgment is false if its object does not exist; a negative judgment, if its object exists.

Likewise, Ajdukiewicz remarks on:

> the popular definition of a true sentence according to which: a sentence is true if there is something in reality corresponding to that sentence, scil. that, which is asserted by the sentence [Ajdukiewicz, 1966, 40].

The "correspondence theory" in these formulations,[4] then, was the basic analysis of the concept of truth in Warsaw while Tarski was a student.

The second tradition involved taking it that the T-sentences express the concept of truth. Wolenski and Murawski attibute a focus on the T-sentences to Czeżowski[5]:

> Truth is a property of special significance. If a certain sentence *A* is true, the sentence *A is true* is also true, if one of them is false, then so is the second: the sentences *A* and *A is true* are equivalent [Czeżowski, 1918, 106] (my translation).

Strictly speaking, though, this passage advocates something like the redundancy theory, discussed and rejected by Kotarbiński [Kotarbiński, 1966, 107ff] in his discussion of the inadequacy of the "verbal" interpretation of truth, and explicitly attributed by Woleński and Murawski to Zygmunt Zawirski on the same page, and not explicitly the idea that the T-sentences are true.[6] However, since distinguishing equivalence from the truth of a biconditional requires consideration of logics for which the deduction theorem fails, Woleński and Murawski's attribution of the view to Czeżowski can be accepted as fair enough.

Tarski himself, of course, attributes the idea that the T-sentences are of fundamental importance in the theory of truth to conversations with Leśniewski and to nobody else [Tarski, 1983a, 154]. His own discussion of truth begins with a contrast between the "classical" and "utilitarian" conceptions that stems from Kotarbiński [Tarski, 1983a, 153]. Kotarbiński favors the classical conception, his objections to the utilitarian conception being standard objections to pragmatist accounts of truth [Kotarbiński, 1966, 106–7].[7] Of the classical conception Kotarbinski writes:

> In the classical interpretation, truly means in accordance with reality...
>
> Let us ... ask what is understood by "accordance with reality". The point is not that a true thought should be a good copy or simile of the thing of which we are thinking, as a painted copy or a photograph is. A brief reflection suffices to recognize the metaphorical nature of such a comparison. A different interpretation of "accordance with reality" is required. We shall confine ourselves to the following: "John thinks truly if and only if John thinks that things are so and so, and things in fact are so and so."
>
> If anyone finds the meaning of that formula insufficiently clear, let him consider the following examples of its application: for instance, the central idea of the Copernican theory is true; it consists in the assertion that the Earth revolves around the Sun; now Copernicus

thought truly, for he thought that the Earth revolves around the Sun, and the Earth does revolve around the Sun [Kotarbiński, 1966, 106–7].

Tarski obviously has this passage in mind in the early pages of "The Concept of Truth", both because he refers to Kotarbiński and because he explicitly makes Kotarbiński's distinction between the "classical" and "utilitarian" conceptions of truth. Tarski has shifted from Kotarbiński's treatment of "thinking truly" to "true" as applied to sentences, but given the Intuitionistic Formalist background this is but a small change; cf. [Kotarbiński, 1966, 105] where the view is that "true" as applied to sentences derives from "truly" as applied to thinking. Given Tarski's remarks at [Tarski, 1983a, 174] he probably agrees, but leaves the matter off the record.

Since Tarski takes himself simply to be showing how to introduce an expression constrained by theorems to express the concept of truth so construed, he accepts Kotarbiński's analysis of the concept wholesale, and likewise simply takes on board the idea, probably from Leśniewski, that the T-sentences somehow express the concept of truth. This is significant because doing both things at once seems to involve the conflation of two ideas that are today thought to be quite opposed: the idea that truth is some kind of correspondence and the basic deflationist idea that truth is somehow analyzed by the T-sentences [David, 1994] [Patterson, 2003]. This of course significantly complicates the question, to which we will return, of whether Tarski was a deflationist. Of more proximate importance here, though, is the simple fact that the T-sentences don't in any obvious way express the "classical" conception of truth common to predecessors like Twardowski and Kotarbiński, yet Tarski treats the two ideas as obviously bearing some close relationship:

Amongst the manifold efforts which the construction of a correct (*korrekt, poprawnej*) definition of truth for the sentences of colloquial language has called forth, perhaps the most natural is the search for a *semantical definition*. By this I mean a definition which we can express in the following words:

(1) *a true sentence is one which says that the state of affairs is so and so, and the state of affairs is so and so*

From the point of view of formal correctness, clarity, and freedom from ambiguity of the expressions occurring in it, the above formulation obviously leaves much to be desired. Nevertheless its intuitive meaning and general intention seem to be quite clear and intelligible. To make this intention more definite, and to give it a correct form, is precisely the task of a semantical definition.

As a starting-point certain sentences of a special kind present themselves which could serve as partial definitions of the truth of a sentence or more correctly as explanations of various concrete turns of speech of the type '*x* is a true sentence'. The general scheme of this kind of sentence can be depicted in the following way:

(2) *x is a true sentence if and only if p.*

In order to obtain concrete definitions we substitute in the place of the symbol '*p*' in this scheme any sentence, and in the place of '*x*' any individual name of this sentence [Tarski, 1983a, 155–6].

Note here that for once Woodger's "correct" is actually correct: the German and Polish are "*korrekt*" and "*poprawny*". This is very important: in the transition from (1) to (2) the semantical definition is not rejected for being intuitively inadequate, but for being formally incorrect. The semantical definition cannot, on its own, serve as a definition of truth without bringing paradox in its wake and, more importantly for our purposes now, it is, though a definition, one that doesn't eliminate meaning-theoretic vocabulary, since it explicitly appeals to the notion of *saying*. So despite its naturalness as a definition of truth—despite, that is, the fact that it does accurately state the content of the concept of truth—it cannot be incorporated directly into a serious deductive theory in metamathematics. Tarski thinks that the semantical definition actually does express the content of the concept of truth; he just doesn't think that it can be directly incorporated into a proper metamathematical theory. His attention thus turns to the second aspect of Polish thought about truth: the T-sentences.

The problem with the passage from (1) to (2) is that (2) doesn't in any obvious way simply say what (1) says. Yet a "semantical definition" is supposed to work out the idea expressed in (1), its "starting point" is the truth of the T-sentences, and Convention T is simply the idea that an intuitively adequate definition—one that expresses the content of the concept of truth, which Tarski explicitly holds is expressed, though not with rigor, by (1)—sums up the T-sentences. One of the main puzzles an interpreter needs to solve here is how the "logical sum" of the T-sentences could be somehow closely related to the "semantical definition". When we are able to do this we will understand how Tarski saw the two strands of Polish thought about truth fitting together.

4.1.3 Semantic concepts in a mathematical theory

After a couple of examples of T-sentences formed both quotationally and with structural–descriptive names of sentences, Tarski notes that paradoxical sentences can be formed in colloquial language. He writes:

> A certain reservation is nonetheless necessary here. Situations are known in which assertions of just this type, in combination with certain other not less intuitively clear premises, lead to obvious contradictions, for example the *antinomy of the liar* [Tarski, 1983a, 157].

He then follows with a simple case of "contingent" paradox, using a sentence that says that the sentence printed in a certain location is not true, where the sentence in question is this sentence itself, and continues:

> The source of this contradiction is easily revealed: in order to construct assertion (β) we have substituted for the symbol 'p' in the scheme (2) an expression which itself contains the term 'true sentence' (which the assertion so obtained ... can no longer serve as a partial definition of truth). Nevertheless no rational ground can be given why such substitutions should be forbidden in principle [Tarski, 1983a, 158].

Notice that at this point Tarski has not suggested somehow quantifying over the variables in (2) to produce a definition, so the antimony of the liar doesn't raise a problem just for an attempted semantical definition of truth for colloquial language. The paradox as formulated calls into question not merely whether truth can be defined by the T-sentences, but whether the T-sentences themselves can even be jointly true. This is important for our understanding of his general remarks, as is his comment that "no rational ground" can be given for forbidding substitutions into the T-schema of sentences themselves that contain the word "true". One might think that avoidance of falsehood and paradox would be rational grounds for such a thing, so the comment must pertain to the "spirit" of everyday language, as he calls it in his general remarks: forbidding sentences that had as their truth conditions that certain sentences be true would not be in the spirit of everyday language, as it would be directly contrary to its presumed universality [Tarski, 1983a, 164].

Setting aside the worry about semantic paradox for the nonce, Tarski next considers how one might generalize the T-sentences into a definition. He considers simply binding the sentential variable in the disquotation schema with a sentential quantifier, producing:

(5) *for all p, 'p' is a true sentence if and only if p* [Tarski, 1983a, 157]

but notes that such a definition wouldn't eliminate "is true" in all contexts. In order to overcome this, the suggested semantic definition is

(6) *for all x, x is a true sentence if and only if, for a certain p, x is identical with 'p' and p* [Tarski, 1983a, 159].

This is then followed by his rather infamous remarks on quotation. Tarski considers two readings of quotation marks. On the first, "quotation-mark names may be treated like single words of a language, and thus like syntactically simple expressions" [Tarski, 1983a, 157]. As such, he maintains, the occurrence of "p" inside of quotation marks isn't bound by the quantifier in (5) and (6), with the results that it is a consequence of (6) that the letter "p" is the only true sentence and that (5) is inconsistent, since the letter "p" will be assigned every truth condition. The remarks here are objectionable on a number of fronts, but to save space I will not comment on the matter; see for instance [Soames, 1999, 86–90].

The other proposed reading of quotation marks is in terms of "quotation functions"

The quotation marks then become independent words belonging to the domain of semantics, approximating in their meaning to the word 'name', and from the syntactical point of view they play the part of functors. But then new complications arise. The sense of the quotation-function and of the quotation marks themselves is not sufficiently clear. In any case such functors are not extensional; there is no doubt that the sentence *"for all p and q, in case (p if and only if q), then 'p' is identical with 'q'"* is in palpable contradiction to the customary way of using quotation marks. For this reason alone definition (6) would be unacceptable to anyone who wishes consistently to avoid intensional functors and is even of the opinion that a deeper analysis shows it to be impossible to give any precise meaning to such functors [Tarski, 1983a, 161].

Furthermore, as Tarski notes, promiscuous use of quotation marks so construed will allow us to reconstruct the liar paradox.

Tarski's worry about intensionality is certainly apposite: the reference of a quotation is no function of any semantic property of what it names. Having considered these two interpretations of quotation marks that he finds unsatisfactory, Tarski gives up and moves on to a perfunctory discussion of the obviously doomed strategy of providing a structural definition of truth of the sort we have encountered before in terms of the closure of some set of sentences under certain transformations; outside of an axiomatic context the procedure is senseless.

§ 1 is advertised as an argument that

> the very possibility of a consistent use of the expression 'true sentence'
> which is in harmony with the laws of logic and the spirit of everyday
> language seems to be very questionable, and consequently the same
> doubt attaches to the possibility of constructing a correct definition
> of this expression [Tarski, 1983a, 165].

This it is, and we will return to the issue in the next chapter. Since on
Tarski's view the T-sentences express the concept of truth as grasped in
the semantical definition, but the T-sentences in colloquial language cannot even be jointly true, Tarski cannot follow the strategy of expressing
the content expressed by the semantical definition by taking the the
instances of (2), the T-sentences, to be one and all assertible sentences of
colloquial language. The idea is then this, that if we do want to express
the concept of truth as grasped in the semantical definition, and if we
are committed to the view that we can do so by making the T-sentences
for the sentences of a language one and all assertible in our theory, this
theory cannot itself be stated in colloquial language; nor can it take
colloquial language as its object language either, since the universality
of colloquial language means that the language of the theory could be
translated into the object language, creating the conditions for the paradox. We thus turn to expressing the concept of truth only in (and for)
non-universal formalized languages [Tarski, 1983a, 167], and the work
of the paper begins in earnest with § 2 and its syntactic treatment of the
language of the calculus of classes.

4.1.4 T-sentences

At the outset of § 3, after noting the weakness of a definition of truth in
terms of provability, Tarski suggests:

> returning to the idea of a semantical definition as in § 1. As we
> know from § 2, to every sentence of the language of the calculus of
> classes there corresponds in the metalanguage not only a name of this
> sentence of the structural–descriptive kind, but also a sentence hav-
> ing the same meaning. For example, corresponding to the sentence
> '$\prod x' \prod x'' A I x' x'' I x'' x'$' is the name '$\bigcap_1 \bigcap_2 (\iota_{1,2} + \iota_{2,1})$' and the sentence
> 'for any classes a and b we have $a \subseteq b$ or $b \subseteq a$'. In order to make
> clear the content of the concept of truth in connexion with some one
> concrete sentence of the language with which we are dealing we can
> apply the same method as was used in § 1 in formulating the sen-
> tences (3) and (4) (cf. p. 156). We take the scheme (2) and replace the

symbol 'x' in it by the name of the given sentence, and 'p' by its translation into the metalanguage. All sentences obtained in this way, e.g.
'$\bigcap_1 \bigcap_2 (\iota_{1,2} + \iota_{2,1})$ *is a true sentence if and only if for any classes a and b we
have $a \subseteq b$ or $b \subseteq a$*', naturally belong to the metalanguage and explain
in a precise way, in accordance with linguistic usage, the meaning of
phrases of the form '*x* is a true sentence' which occur in them. Not
much more is to be demanded of a general definition of true sentence
than that it should satisfy the usual conditions of methodological correctness and include all partial definitions of this type as special cases;
that it should be, so to speak, their logical product [Tarski, 1983a, 187].

This set of ideas provides for truth the clear conception of what is to be
accomplished that was lacking for definability in "On Definable Sets",
one precondition of which was sufficient resources in the metalanguage
for a proper treatment of use and mention of sentences in the object
language and their translations into the metalanguage. The content of
the concept of truth is, according to the "semantical definition", the
property of saying that so-and-so when things are so-and-so. The task is
to craft a deductive theory that forces an expression "∈ *Tr*" to have this
content as its meaning, and the task will be taken to be accomplished
when the theory is made to imply the T-sentences for "∈ *Tr*".

"∈ *Tr*" will express the concept of truth just in case it is associated
by the conventions of the metalanguage with the content expressed by
the semantical definition. However, unlike the case of consequence as
construed around 1930, where Tarski was able simply to require that
theorems that state the conception associated with the concept of consequence be provable in the metatheory, no similar requirement will do in
the case of the semantic concept of truth, since the semantical definition
cannot be part of a formally correct metamathematical theory.

Thus means need to be found to constrain the intuitive interpretation
of "∈ *Tr*" down to the point where it expresses truth analyzed according
to the semantic definition while guarding against the problems direct
incorporation of the semantical definition into a theory would bring in
its wake. Since the switch to non-universal formalized languages allows
the T-sentences for the object language to be consistent, we can make
use of them. But the question remains of how having the T-sentences
as theorems makes "∈ *Tr*" express the concept of truth as grasped in
the semantical definition. This was the interpretive puzzle about the
semantical definition and the T-sentences noted above.

The genius of Convention T is that incorporating the T-sentences into
the metatheory forces the interpretation of "∈ *Tr*" to accord with the

semantical definition, while leaving the definition itself unstated. This, I propose, is the connection between the semantical definition and the T-sentences, the two elements of Polish thought about truth that Tarski brings together. To put it another way, if the T-sentences are theorems, "$\in Tr$" expresses the irreducibly semantic content of the concept of truth, the content expressed in the "semantical definition", and if the T-sentences are theorems because the metatheory has been extended by a certain definition, we thereby force a theory with only mathematical primitives to express a semantic concept, thereby meeting the eliminative goal stated at [Tarski, 1983a, 154] that undefined semantic terms will not appear within the metatheory.

Here is how the trick works. When we put the elements together, what we find is that an expression for which the T-sentences are theorems is, given the conventions governing the object language and the metalanguage, forced to express the content of the concept of truth as construed in the semantical definition. Since the language of the metatheory has intuitive meaning just as much as does the object language (see the discussion of translation below in § 4.3.1), each T-sentence is meaningful according, as Kotarbiński would say, to the adopted conventions or, as Tarski has it in his change to the German (§ 1.4.3) in accordance with linguistic usage. In particular, a T-sentence like:

$\bigcap_1 \bigcap_2 (\iota_{1,2} + \iota_{2,1})$ is true if and only if for any classes a and b we have $a \subseteq b$ or $b \subseteq a$

means that $\bigcap_1 \bigcap_2 (\iota_{1,2} + \iota_{2,1})$ is true if and only if for any classes a and b we have $a \subseteq b$ or $b \subseteq a$. Now, though the T-sentence itself says nothing of its own *saying* that something is the case, a party to the conventions governing the metalanguage will recognize that in fact the T-sentence does say (in Kotarbiński's terms, state indirectly) exactly that.

Likewise, a party to the conventions governing the object language will recognize that $\bigcap_1 \bigcap_2 (\iota_{1,2} + \iota_{2,1})$ says that for any classes a and b we have $a \subseteq b$ or $b \subseteq a$. Hence, a party to both sets of conventions will recognize that what the T-sentence says is that $\bigcap_1 \bigcap_2 (\iota_{1,2} + \iota_{2,1})$ is true if and only if so-and-so, where so-and-so is precisely what $\bigcap_1 \bigcap_2 (\iota_{1,2} + \iota_{2,1})$ says. The concept "is true" is thereby constrained by the theoremhood of the T-sentence to apply to $\bigcap_1 \bigcap_2 (\iota_{1,2} + \iota_{2,1})$ if and only if what $\bigcap_1 \bigcap_2 (\iota_{1,2} + \iota_{2,1})$ says is the case is in fact the case, and someone who accepts the conventions governing both object and metalanguage will thereby accept this. As a result, since the T-sentence is a theorem, "$\in Tr$" as applied to $\bigcap_1 \bigcap_2 (\iota_{1,2} + \iota_{2,1})$ has to express the content of the concept of truth.

Getting the hen-scratches out of the picture may help the reader. Consider the French sentence

La neige est blanche.

and its English T-sentence, with a dummy letter substituted for "true":

"La neige est blanche" is F if and only if snow is white.

Consider now a party to the conventions of French who is also a party to the conventions of English who begins by associating no intuitive content with F. Such a person will accept that

"La neige est blanche" is associated by the conventions of French with the thought that snow is white.

since accepting this is part of her being party to the conventions of French. Equivalently, she will accept that

By the conventions of French "La neige est blanche" says that snow is white.

Now suppose this person considers the modified English T-sentence, wondering what she would be committed to by it. What she does know is that were she to accept it, she would be accepting that "La neige est blanche" falls within the extension of F if and only if snow is white. She also knows that "La neige est blanche" says that snow is white. So she knows that "La neige est blanche" says that the state of affairs is that snow is white, and she knows that the T-sentence says that "La neige est blanche" is F if and only if the state of affairs is that snow is white. So she knows that, when it comes to "La neige est blanche", there is a state of affairs that so-and-so such that "La neige est blanche" says that the state of affairs is so-and-so and (by the modified T-sentence) it is F if and only if the state of affairs is so-and-so. Hence, if she does possess the concept of truth as analyzed by the semantical definition, she will recognize that "La neige est blanche" falls within the extension of F just in case it has the property that is the content of the concept of truth, namely, saying that the state of affairs is so-and-so when the state of affairs is so-and-so. She may therefore rightly conclude that at least when restricted to "La neige est blanche", a sentence falls within the extension of F just in case it falls under the concept of truth. Similar thought about T-sentences for the rest of the language will lead to the conclusion that F applies to a sentence just in case it says that the state of affairs is so-and-so, when so-and-so. F is thereby constrained to express the concept of truth as analyzed by the semantical definition, yet none of the problematic aspects of the semantical definition have been explicitly incorporated into the metatheory for French stated in English: the notions of "saying", "state of affairs" and the schematic "so-and-so" all remain within the mind of one who contemplates the metatheory and the object language while accepting their Intuitionistic Formalist meaning-determining conventions. We thus accept that the semantical definition is correct and force the concept so-analyzed to be expressed by "∈ *Tr*" without the formal incorrectness that would come from direct inclusion of the semantical definition in the metatheory.

A point should be entered now on the respect in which Tarski aims to "reduce" semantic concepts to other concepts [Tarski, 1983a, 153]. The reading of this that comes most easily to the mind of the contemporary reader is that within Tarski's theories, semantic concepts will be explicitly defined in other terms, and that *thereby* it will somehow be shown that the *concept* of truth can be analyzed in non-semantic terms. This is not what Tarski is up to. Precisely because he has no intention of providing an analysis of the concept of truth at all, Tarski *a fortiori* has no intention of giving a reductive analysis of the concept. Tarski simply assumes Kotarbiński's analysis, and Kotarbiński's analysis doesn't eliminate "says". For all Tarski ever does or shows, the target concept itself is irreducibly semantic; indeed, since the "semantical definition" is the "most natural", it seems that Tarski thought as much. His goal isn't to get rid of this; his goal, rather, is to show that a properly structured mathematical deductive theory can constrain one of its terms to *express* a semantic concept. It is to this end that Convention T is directed.

We can make a general point about Intuitionistic Formalist metamathematics as Tarski develops it here. There is nothing intrinsically reductive about Tarskian metamathematics. The aim isn't to show that semantic (or other metatheoretic) concepts can be defined away, that they are safe for mathematics or physics, or anything else of the sort.[8] The aim is to explore the extent to which the properties of sentences and deductive theories attributed in intuitive metatheoretic thought can be treated of in theories in which the forms of expressions go proxy for these properties. Convention T and definitions that are intuitively adequate according to it are the crowning achievement of this program: given the conventions of the object- and metalanguages, the theoremhood of the T-sentences gives a deductive theory a form that constrains one of its expressions to express a target semantic concept. (In the metametalanguage, of couse, we don't specify the relevant form entirely syntactically for it is here that intuitive meanings and conventions of language are appealed to in Convention T's notion of translation.)

Tarski states Convention T as follows:

> Convention T. *A formally correct definition of the symbol 'Tr', formulated in the metalanguage, will be called an* adequate definition of truth *if it has the following consequences*:
> (α) *all sentences which are obtained from the expression 'x ∈ Tr if and only if p' by substituting for the symbol 'x' a structural–descriptive name of any sentence of the language in question and for the symbol 'p' the expression which forms the translation of this sentence into the metalanguage;*
> (β) *the sentence 'for any x, if x ∈ Tr then x ∈ S' (in other words, 'Tr ⊆ S')*
> [Tarski, 1983a, 187-8].

Convention T fixes the problem we found in ODS: it states, in the metametalanguage, a criterion of adequacy for a defined expression of a syntactic deductive theory for some object language to express the concept of being a true sentence of that language. The relativization of the concept expressed to the object-language is left implicit in the same ways as it is left implicit in all of Tarski's other definitions of linguistic notions, e.g. his definitions of sentencehood in CTFL or in the early papers on the consequence relation. Again, this is not an analysis of the concept of truth, nor is it a criterion of adequacy on the analysis of this concept. For Tarski the philosophers, in particular Kotarbiński and Aristotle, have taken care of that. It is a criterion of adequacy on an expression's being constrained by the structure of a deductive theory to *express* the concept of truth so analyzed.

4.2 Tarski's definitions

People often speak of "Tarski's definition of truth". This is a misnomer. The definition set out in § 3 is a formal definition of "∈ *Tr*" for the specific language of a specific deductive theory, namely the first-order theory of the subset relation expressed in Polish notation with certain specific axioms. The remainder of § 4 then concerns how the methods employed in setting out that definition can be modified and extended to construct similar definitions for other deductive theories in other formal languages. None of the proposed definitions are themselves analyses of the concept of truth; this analysis, borrowed from Kotarbiński, takes place in the metametalanguage as the motivation for Convention T just discussed.

4.2.1 Truth for the language of the calculus of classes

We can afford to be brief and informal in discussing Tarski's example itself. The labor of § 2 goes into development of string theory, formal syntax and proof-theoretic consequence, the only notion of consequence at work in 1931 [Tarski, 1983a, 166], for Tarski's first-order theory of the subset relation. Definition 22 defines satisfaction for the language in the familiar way. Since the only non-logical primitive is $\subseteq$, represented as an *'I'* preceding two variables, an atomic open sentence Ix_ix_j is satisfied by a sequence iff the ith element is a subset of the jth; negation, disjunction and the quantifier get the now-obvious treatment. It follows from this definition that sentences with no free variables are satisfied either by all sequences or by none. Definition 23 then says that x is a true sentence if and only if every sequence of classes satisfies it.[9]

What closes the gap between mathematics and semantics that was left open in ODS is care about use and mention facilitated by the structural–descriptive expressions for the object language, plus the enumerative definition of satisfaction for atomic predicates of the object language (of which there is only one in the case at hand). Below we will consider the extent to which this really closes the gap between semantics and mathematics in a way that would meet various demands.

Tarski then sets out to show that the definition satisfies Convention T. He prescinds from doing this in detail on the grounds that doing it with rigor would require the formalization of the metatheory and a proof in the metametatheory that Def. 23, against the background theory and preceding definitions does in fact imply all of the T-sentences for the language of the calculus of classes [Tarski, 1983a, 195]. Rather we are to convince ourselves of the intuitive adequacy of Def. 23 given the background theory and prior definitions by "the empirical method".[10] Since it is the case that the definition plus the prior theory and definitions imply the T-sentences, we can move on.

Tarski also points out that "some characteristic general theorems" can be derived, in particular that no sentence and its negation are true and that every sentence or its negation is true, that consequences (in the sense of § 2) of true sentences are true, and so on [Tarski, 1983a, 197–8]. These theorems are required to "fix the conviction of the intuitive adequacy" (*"sachlichen Richtigkeit"*, *"merytorycznej trafności"*) of Def. 23 because Tarski holds that "no definition of true sentence which is in agreement with the ordinary use of language should have any consequences which contradict the principle of the excluded middle" [Tarski, 1983a, 186]—and, presumably, similarly for the other points. This is underemphasized in Tarski's remarks on the idea behind a semantical definition in § 1, but the idea has to be that for any sentence that says something, what it says is either the case or not. Of course, this ultimately assumes the law of the excluded middle in the metametalanguage, but this sort of consideration doesn't bother Tarski—after all, he is happy to enshrine as Theorem 5 of § 3 the claim that the theorems of the object theory are true, which is just a consequence of the fact that the theory stated in the object-language is a sub-theory of the mathematical theory accepted in the metalanguage. Tarski remarks on this sort of thing, with reference to [Ajdukiewicz, 1966], in connection with proving the consistency of the object theory, that this doesn't make the proofs of the theorems circular, though they fail to "add much to our knowledge" [Tarski, 1983a, 237].

Unfortunately this discussion makes clear that Tarski's treatment of intuitive adequacy is inconsistent. Convention T itself states that a theory's implying the T-sentences is sufficient as well as necessary for intuitive adequacy. But [Tarski, 1983a, 186] states that "no definition of true sentence which is in agreement with the ordinary use of language should have any consequences which contradict the principle of excluded middle" and Tarski holds that proving generalized excluded middle (Theorem 2) is something done "in order to fix the conviction of the intuitive adequacy (*"sachlichen Richtigkeit"*, *"merytorycznej trafności"*) of the definition" [Tarski, 1983a, 197] (corrected). So Convention T and other passages make implying the T-sentences sufficient for intuitive adequacy, while 186 and 197 seem to imply that it isn't, since implying various generalizations such as non-contradiction and excluded middle are necessary. I don't get the impression that Tarski noticed this problem, but since both Convention T and the later claim that axiomatic truth theory is intuitively adequate, when Tarski explicitly notes as a weakness of it that it doesn't imply the desired generalizations, indicate that implying the T-sentences is the heart of the matter. Moreover, the provability, say, of excluded middle obviously depends on the logic in the metatheory and doesn't really have anything to do with the conception of truth. Since a terminological distinction between, say, "intuitive adequacy" and "complete intuitive adequacy" or something of the sort would clear things up I will not pursue the matter further.

4.2.2   Higher order and polyadicity

With one example worked out in detail, Tarski turns in § 4 to a general discussion of the application of his method for defining truth. Much of the discussion is, again, familiar enough: add matter sufficient for formal syntax plus a translation of the object language to some sufficient background logic, and then proceed to the definition of satisfaction [Tarski, 1983a, 210–11]. What requires further discussion in the way in which satisfaction needs to be handled for languages of varying expressive power.

The issues here are both technical and, for Tarski, philosophical. The technical issues mainly concern mathematical bookkeeping: when we allow for variables of different orders or relations of varying adicity we need need to keep track of them in our sequences. Order itself is handled rather easily; the introduction of second order variables simply requires "two-rowed sequences", one row comprising objects over which first-order variables range and the other objects over which second-order variables range [Tarski, 1983a, 237]. Higher but finite order goes through with the addition of more rows. Variable adicity causes more trouble. If it isn't bounded above, we don't have a finite number of rows in our many-rowed sequences anymore, since a many rowed sequence needs a first row of objects, a second row of sets, that is, monadic relations, a third row of dyadic relations, a fourth row of triadic relations, and so on [Tarski, 1983a, 232]. This problem can be solved, however, by taking satisfaction to be a relation between open sentences, infinite sequences of objects, and infinite sequences of sets of finite sequences of objects [Tarski, 1983a, 234].

In all these cases Tarski also notes that by exploiting one-one correlations between objects and certain relations (or the sets of sequences they semantically define) we can get by in the cases discussed with a single infinite sequence of appropriately selected objects [Tarski, 1983a, 228, 235]. This is what he calls 'the method of semantical unification of the variables' [Tarski, 1983a, 227]. A simple example of the method is provided at [Tarski, 1983a, 228]: since every individual x can be one-one correlated with the relation that holds between y and z if and only if $y = x$ and $z = x$, we can simply take first-order variables to range over these relations. Nothing more than a little arithmetic is required to keep the ranges of the first- and second-order variables distinguished within the sequences ([Tarski, 1983a, 229]) and we can in this manner take satisfaction again to be a relation between an open sentence and a single infinite sequence. Tarski notes one hitch that will matter later: the unifying category cannot be of lower order than the highest-order variable in the language [Tarski, 1983a, 230].

The philosophical issues are of more relevance to us. They center on what Tarski calls "the theory of semantical categories", which he credits to Husserl, with reference also to Leśniewski and Ajdukiewicz (whose version was later published as [Ajdukiewicz, 1935]), which he relates to the type theory of *Principia Mathematica* and the simplified type theory (STT) of Chwistek and Carnap's *Abriss*.[11] Unfortunately no written presentation of it by Leśniewski exists, though it gets some mention at [Leśniewski, 1992e, 421, 477–8]. However, Ajdukiewicz presents a version of it in [Ajdukiewicz, 1978d] and the ideas may be familiar to the reader from [Lewis, 1970]. Applied to a language the view assigns a semantic category to every expression, taking some two categories (*sentence* and something else, usually *name*) as basic, and expressing other categories as functions from n-tuples of categories to some category (e.g. a one place predicate is an S/N, an adverb is an $(S/N)/(S/N)$ and so on). Tarski comments on its significance:

> Whilst the theory of types was thought of chiefly as a kind of prophylactic to guard the deductive sciences against possible antinomies, the theory of semantical categories penetrates so deeply into our fundamental intuitions (*fundamentalen Intuitionen, podstawowe intuicje*) regarding the meaningfulness of expressions, that it is scarcely possible to imagine a scientific language in which the sentences have a clear intuitive meaning (*inhaltlichen Sinn, intuicyjny sens*) but the structure of which cannot be brought into harmony with the above theory [Tarski, 1983a, 215].

(Compare Leśniewski on the concept of a semantical category: "Frankly, I would still today feel obliged to accept this concept even if there were no antinomies at all" [Leśniewski, 1992e, 421].) The categories are equivalence classes of expressions intersubstitutable within sentences (open or closed) preserving sentencehood. The philosophical issues depend, for Tarski, on the view we have already mentioned that intuitive meaning is tied to semantical category. With respect to a formulation of second-order quantification over two-termed relations, he writes:

> To obtain a correct and adequate (*korrekte und richtige, poprowna i trafna*) definition of satisfaction in connexion with the language we are considering we must first extend our knowledge of the concept. In the first stage of operating with it we spoke of the satisfaction of a sentential function by one, two, three objects, and so on, according to the number of free variables occurring in the given function (cf. pp. 189 ff.). From the semantical standpoint the concept of satisfaction had there a strongly ambiguous character; it included relations in which the number of terms was diverse ... Strictly speaking we were dealing not with *one* concept, but with an infinite number of analogous concepts, belonging to different semantical categories. If we had formalized the metalanguage it would have been necessary to use

infinitely many distinct terms instead of the *one* term 'satisfies'. The semantical ambiguity of the concept increases still more when we pass to languages of more complicated logical structure ... the semantical category of a relation not only depends on the number of domains, i.e. the number of terms standing in the relation to one another, but also on the categories of these domains ... To functions which belong to two distinct types two semantically different categories always correspond [Tarski, 1983a, 224–5, my correction].

One way the point emerges here is this: a fully formalized treatment of a language that conforms to the requirements of the theory of semantical categories requires a metalanguage that also so conforms, but then the facts of the case will mean that no expression can express "the" satisfaction relation that obtains between both, say, first-order one-place open sentences and sequences of objects and second-order sentences open on a two-termed relational expression and sequences of ordered pairs.

The *concept* expressed by an occurrence of "satisfies" is determined in part by the number and order of the expressions required to form a sentence with it, so that in our two cases we have the forms $sat(x, Y)$ and $sat(x, y)$. Furthermore, for Tarski, the theory of semantical categories is to constrain "independent meaning" (*selbständige Bedeutung, samodzielnego sensu*), which is allied with his notions of concept and intuitive meaning. His treatment of semantical category in the text, he notes, includes expressions without "independent" meaning, such as "it is not the case that", "For all x" and "either x is a subset of y or" [Tarski, 1983a, 217]. A proper treatment would constrain the theory down to cover expressions only with independent meanings, which means that the theory of semantical categories, properly conceived and fully developed, is a theory of the conditions under which expressions with independent meaning can compose to form a whole with independent meaning. These remarks make clear that the theory of semantical categories is ultimately concerned with our capacity to attach intuitive meanings to expressions, in accord with "our fundamental intuitions regarding the meaningfulness of expressions" [Tarski, 1983a, 215]. But if the theory thus constrains these intuitive meanings, we get the conclusion that there is no single concept of satisfaction.

We can see in this passage that even in 1931 the constraints of STT are beginning to chafe. As Tarski notes in several places in CTFL and elsewhere (e.g. [Tarski, 1983k, 310 n. 1]) many of his formulae are actually schematic over the simple type-theoretic hierarchy. When he encountered this problem, he had no qualms about engaging in "systematically ambiguous" schematic generalization over types, yet Tarski and his readers treat the resulting formulations as perfectly intelligible. It is in such moments that we can see the seeds of Tarski's shift to a more tolerant (in Carnap's sense) attitude toward logical systems, the relaxation of the restrictions of STT and the theory of semantical categories in the 1935 Postscript (§ 6.1) and his later move to first-order logic for many purposes. We will return to these issues below. CTFL thus bears an interesting and as far as I know previously unnoticed resemblance to Wittgenstein's *Tractatus*: the body of the work consists substantially of expressions that are nonsense by the work's own standards.

4.2.3   Domain relativization and consequence

Tarski, both in his specific example of the language of the calculus of classes and also in the general discussion of § 4 also treats a topic that looks much more to be a forerunner of contemporary model theory: "the notion of *correct* or *true sentence in an individual domain a*" [Tarski, 1983a, 199]. The modifications to the definitions of truth and satisfaction are straightforwrard: we simply build our infinite sequences from objects in the specified set and everything else runs as before [Tarski, 1983a, 200]. Tarski also makes the point that the deductive theory at issue can't distinguish domains of the same cardinality, so that the only important notion is that of truth of a sentence in a domain of cardinality k (Lemma G [Tarski, 1983a, 202]) and he also defines truth in all domains [Tarski, 1983a, 201]. Tarski sets out a presentation of the Löwenheim–Skolem theorem [Tarski, 1983a, 204–5] and the discussion culminates with an "accidental" structural definition of truth (Th. 28) for the language of the calculus of classes. It is sometimes worried that Tarski only imagines restricting the domain and that his definitions don't take into account truth in domains that include objects that don't exist. Since Tarski's interests were entirely mathematical and his theories extensional this is no surprise; moreover, as McGee suggests [McGee, 1992, 278] as long as, as is plausible, any possible structure is isomorphic to some pure set-theoretic structure, there is no harm in this.[12]

CTFL remains old-fashioned in its understanding of logical consequence. By 1931 the truth-definition had been born, but with respect to consequence nothing had changed. Definitions 15 and 16 in § 2 simply work out the idea that "among the consequences of a given class of sentences we include first all the sentences belonging to this class, and all the sentences which can be obtained from these by applying, an arbitrary number of times, the four operations of *substitution, detachment,* and *insertion* and *deletion of the universal quantifier* [Tarski,

1983a, 181]. This, as discussed in the introduction, ties the notion of consequence to a set of intuitively valid rules of inference and hence doesn't provide what is of the essence of logic as we know it today: the interplay between proof-theoretic and semantic conceptions of consequence. As Tarski notes, it also doesn't capture the consequence relation intuitively. This issue will loom large in Chapter 7.

Since Tarski's semantics can be put to this use, as he himself put it in 1935, we see here, as we did with his treatment of Padoa's method, what looks like another blind spot for Tarski in the early 1930s. As with Padoa's method, however, what this really shows isn't that Tarski wasn't seeing something that he was somehow looking for, but that he wasn't looking for what today we take it was staring him in the face. In 1931 Tarski had no intention of replacing the Leśniewskian idea that the purpose of a deductive theory is to express intuitively grasped thoughts and the intuitive inferential connections between them with semantics; semantics was, rather, a tool for producing definitions that introduced into deductive theories so construed terms that expressed metamathematical concepts.

4.3 Evaluating Tarski's account

Tarski occupies an intriguing place in philosophical mythology. Although he is universally acknowledged as important, all of his substantive philosophical views are widely agreed to be failures. In this section we will debunk a number of familiar criticisms of what is known generally as "Tarski's Theory of Truth". As we will see, the main theme will be the confusion on which a definition constructed along Tarski's lines is itself somehow Tarski's "theory" of truth.

4.3.1 Familiar questions

I begin with a number of familiar questions and false assumptions about Tarski's views that are of lesser philosophical import than the issues to be discussed below. First and perhaps foremost: Convention T is not, in the first instance, a criterion of extensional adequacy, as is very often claimed.[13] It is, rather, a statement of the conditions under which "$\in Tr$" will be so constrained by the structure of the metatheory that it expresses the intuitive concept of truth (relative to the object language with its meaning-determining conventions associating thoughts with sentences). Confusion on this point is universal: nearly every treatment of Tarski's views that one can find in print treats "material" as meaning "extensional". Extensional adequacy follows from the theoremhood of the T-sentences (assuming that the metatheory is true), but the point of Convention T is not merely to introduce a predicate with the right extension.

In addition to the plain obviousness of the point that it could always turn out, one way or another, that the set of truths of a language happened to be the extension of some extraneous predicate, Tarski's discussion of Th. 28 makes clear that he appreciated the point. Theorem 28 states that, for the language of the calculus of classes, $x \in Tr$ if and only if x follows from the usual axioms, plus the claim ("sentence α") that every non-empty class has a singleton class as a subclass and the negations of all sentences γ_l stating that there exist exactly l things for all natural l. As Tarski notes [Tarski, 1983a, 208], Theorem 28 could just as well be a definition of "$\in Tr$". But Th. 28, treated as a definition on its own, isn't intuitively adequate: adding just it (and the definitions it depends on) to the "morphology of language" for the language of the calculus of classes would yield an extensionally adequate definition of "Tr" but not one that would allow the proof of the T-sentences. Clearly, then, since Tarski explicitly notes the matter, he doesn't equate extensional and intuitive adequacy. (He also notes that the provability of Th. 28 depends on "specific peculiarities" of the language of the calculus of classes as well as the "strong existential assumptions" adopted in the metatheory. The accidental case of a structural characterization of the true sentences afforded by Th. 28 in no way undermines the point that in general the set of truths of a language can't be characterized structurally.)

Moving on, we can now settle the famously vexed question of what Tarski thinks "translation" between object and metalanguage sentences is. One sentence translates another just in case the conventions governing both associate the same thought with them and, more generally, one expression translates another when the conventions associate them with the same thought-constituents. At [Tarski, 1983a, 166] it is clear that the expressions of the object language are meaningful in that "we always ascribe concrete and, for us, intelligible meanings" to them, while at [Tarski, 1983a, 170] the translation of object language expressions by those of the metalanguage is due to the fact that they "have the same meaning". For Tarski there simply cannot be any question about whether one sentence translates another, because there simply cannot be any question whether the thoughts associated with two sentences by the conventions of their languages are the same or not.

A related objection that is often heard is that in Convention T Tarski somehow breaks his promise "not to make use of any semantical concept", since he appeals to the notion of translation. Indeed, it is sometimes suggested that the whole account is implicitly circular if, in turn, two sentences translate one another only if they share their truth conditions. Two replies are in order here. First, the notion of translation is tied to Tarski's account of intuitive meaning, not to the notion of a truth-condition, so that translation isn't in the relevant sense "semantical" [Raatikainen, 2008, 254]. Second, even if translation were understood by Tarski in terms of truth-conditions the account wouldn't be circular; indeed, the account wouldn't be circular if Tarski had simply said that a good definition of truth for a given object language was one that implied all of the *true* T-sentences of the object language. Stating a criterion of adequacy on a definition in a metalanguage using the defined term doesn't render the definition in the object language circular and, in fact, as I have stressed elsewhere, properly understood no evaluation of a usage-reporting definition as correct can proceed without deploying the defined term ([Patterson, 2007], [Patterson, 2008b]).

Next we come to one of the central points of contention in the literature: the fact that Tarski's definitions apply to only one language at a time:

> The extension of the concept to be defined depends in an essential way on the particular language under consideration. The same expression can, in one language, be a true statement, in another a false one or a meaningless expression. There will be no question at all here of giving a single definition of the term. The problem which interests us will be split into a series of separate problems each relating to a single language [Tarski, 1983a, 153].

This issue has already been addressed in Chapter 2: on Tarski's conception of linguistic concepts they have application conditions and hence extensions only relative to a language with its meaning-determining conventions. Tarski suppresses the relativization to a language because the notion of being "in" a language doesn't admit of a metamathematical treatment. The aim of metamathematics is to treat linguistic notions, in the present case semantic ones, only insofar as the forms of sentences, as determined by formation rules, can stand in for these notions. Tarski is willing to speak of "the" concept of truth because he thinks there is only one such concept, which applies to a sentence only relative to the conventions of its language. These matters, however, not being mathematically treatable, retreat into the metametalinguistic commentary, to be kept in mind in judging whether the metatheory's treatment of the object language is intuitively adequate.

It is sometimes wondered why Tarski calls Convention T a "convention". Various hypotheses are floated about this. Our discussion of Kotarbiński and Leśniewski makes the matter clear: it is through the conventions of language that expressions get their meaning. Convention T is a *convention* like any other in this respect: it states a way of using a certain expression that suits it to express a certain intuitive meaning. Recall here that Leśniewski is at points explicit that a definition is a kind of convention for the use of a term.

Tarski's wording it with "if" has occasioned speculation to the effect that he intends Convention T to state only a sufficient condition, but this is squelched by his habitual use of "if", often with help of phrases like "will be called", to express definitions, and his explicit endorsement of this practice in the logic textbook at [Tarski, 1995, 36].[14] Footnote 1 at [Tarski, 1983a, 188] also makes clear that Convention T is supposed to amount to a definition, which requires necessity as well as sufficiency.[15] (The English wrongly states it to be a definition of the "meta-theory"; the German correctly has "*Meta-Metawissenschaft*" [Tarski, 1986a, 99].)

Another issue concerns the claim that is sometimes made (e.g. [Etchemendy, 1988, 55]) that although, supposedly, Tarski requires that a definition of truth imply the T-sentences, his own definitions don't in fact imply the T-sentences, since other resources of the metatheory are required to prove them. What we can now say is that for an Intuitionistic Formalist the fundamental notion of intuitive adequacy applies to an expression–theory pair, not to a definition. What is intuitively adequate for Tarski is, in the first instance, the "linguistic usage" imposed upon a term by the deductive structure of an axiomatic theory. If the usage is right, the role of the term in the theory is intuitively adequate. Derivatively, a definition can be adequate relative to a theory if, when added to that theory, it renders it intuitively adequate. ESS is explicit on the matter:

> Now if we succeed in introducing the term 'true' into the metalanguage in such a way that every statement of the form discussed can be proved on the basis of the axioms and rules of inference of the metalanguage, then we shall say that the way of using the concept of truth which has thus been established is *materially adequate*. In particular, if we succeed in introducing such a concept of truth by means of a definition, then we shall also say that the corresponding definition is materially adequate [Tarski, 1983b, 404].

Tarski thus isn't confused in the way suggested; he realizes that the definitions don't imply the T-sentences all by themselves, and he asks for no more than that they do so in conjunction with the rest of the theory.[16]

Interpreters sometimes make much of Tarski's use of structural descriptive names for sentences, suggesting often that they serve some epistemic function, e.g., making it the case that a sentence can always be "recovered" from its name. This is a red herring. The reason structural descriptive names are needed is simply that without them there would be no way to state the recursion clauses in the definition of satisfaction. Aside from that need Tarski is perfectly happy with any sort of way of naming a sentence. This ought to have been obvious from the mere fact that for a finite language a list-like definition as suggested at [Tarski, 1983a, 188] could perfectly well use unstructured names.

4.3.2 Tarskian definitions and Tarski's "theory"

A large family of standard criticisms derives from the combination of the fact that a given Tarskian definition of truth applies only to a single language as it is actually interpreted with the idea that a particular Tarskian definition itself is somehow Tarski's "theory" of truth. A few classic examples are [Black, 1948, 60], Davidson's claim that "Nothing in Tarski's truth definitions hints at what it is that these definitions have in common" [Davidson, 1990, 288], and Putnam's famous claim that "Tarski's theory fails as badly as it is possible for an account to fail" [Putnam, 1994, 333]. Listing the various objections of this variety, we have the following:[17]

1. Tarski's account doesn't tell us what to do when a new term is added to a language for which we have given a truth definition [Black, 1948, 57–8], [Davidson, 1990, 287].
2. Tarski's account tells us that we need a completely new definition of truth when a new term is added to a language for which we have given a truth definition [Black, 1948, 57–8].
3. Tarski's account doesn't tell us what predicates defined in his way have in common [Black, 1948, 60] [Davidson, 1990, 288, 296].
4. Tarski's account implies that "snow is white" would be true even if "snow" meant what "coal" currently means [Putnam, 1994, 333].
5. Tarski's account implies that "'Snow is white' is true in L if an only if snow is white" is a logical or mathematical truth [Putnam, 1994, 333], [Etchemendy, 1988, 57], [Heck, 1997, 537], [Soames, 1984, 422–4], [Davidson, 1990, 288].

All five objections take Tarski's "theory" of truth somehow to be expressed in his definitions. In addition the fourth understands Tarski's definitions as necessary truths and the fifth takes it that a definition of truth in a mathematical or logical theory is itself a mathematical or logical truth.

The common theme can be dealt with easily: Tarski's definitions are not his "theory" of truth. Tarski has a theory of truth: the "semantical definition" of [Tarski, 1983a, 155] hailing from Kotarbiński and other Polish predecessors and ultimately, as they saw it, from Aristotle. If it had been possible to introduce the semantical definition into a meta-mathematical theory in a formally correct way that would have been the end of the matter. This is why, as far as the concept of truth itself goes, Tarski simply defers to the philosophers.

Tarski's project was not to provide a "theory of truth" in the sense of a conceptual analysis or something like a metaphysical account of the nature of truth. It was to provide, in accord with Intuitionistic Formalism, (i) an account of the conditions under which a term of a deductive theory has a role that constrains it to express the concept of truth, and (ii) the means of introducing terms with such roles, perhaps while meeting further desiderata—in particular that the terms be introduced via explicit definition so as to guarantee relative consistency. Tarski's account of (i) is that "$\in Tr$" expresses the concept of truth as applied to the object language if and only if all T-sentences formed with it are theorems. His account of (ii) is the method of recursion on satisfaction. But, as we have seen in detail above, no more than in any other case (e.g. in the definition of substitution of a variable at [Tarski, 1983a, 180]) does Tarski think that a particular definition of "$\in Tr$" states an analysis of the concept of truth. Indeed, the definition doesn't even state an analysis of the concept of truth restricted to the object language at hand, since for Tarski the only constraint on a definition intended to extend a theory to one containing a term expressing a target concept is that the resultant theory have theorems that constrain the interpretation of the defined term down to the target concept. Tarski's definitions are neither analyses of the concept of truth nor "theories of truth".

It does follow from Tarski's way of proceeding that if a new term is added to a language the definition of truth needs to be altered. If the definition were intended as a conceptual analysis, this would be bizarre, just as commentators who assume that the definition is an analysis have taken it to be. But since the definition isn't a conceptual analysis, the point simply that recursion by satisfaction needs to begin from lexical base clauses.

Is it, however, a failing of Tarski's account that it tells us nothing about how to extend a definition when the object language is extended with a new expression? The answer here is that Tarski's account, taken as a whole, tells us exactly as much as it should tell us, neither more nor less. What Tarski's account tells us is that the new definition for the extended object language has to be such that, when added to formal syntax (which, note, likewise needs to be extended in a way not indicated by the prior syntactic theory) the T-sentences for the extended language become theorems. What the account doesn't tell us is exactly how to do this. This is exactly right, because the account of truth *shouldn't* tell us how to do this. If what we add is an expression that belongs to an already non-empty category of expressions—e.g. a new name alongside those already in the language—what Tarski tells us nothing about is what the denotation of this name is. This is exactly as it should be: why on Earth think that a definition of truth should imply why names refer to what? The point is even more obvious if what we add is a new type of expression (e.g. a two-place functor such as "the son of x and y") previously lacking. Neither the account of the concept of truth nor the account of the conditions under which an expression of a deductive theory expresses it should ever have been expected to tell us anything of the sort.

Tarski's account of truth does, contrary to objection 3, tell us what the defined predicates have in common: they all express the intuitive concept of truth by having the the intuitive contents they express constrained by the theoremhood of the T-sentences. One who knows the conventions assigning meanings to object- and metalanguage sentences thereby recognizes that "$\in Tr$" expresses the concept of truth as analyzed in the semantical definition. Only thinking that a particular definition is somehow "the" theory espoused by Tarski could make it look otherwise. Davidson is absolutely right that there is something more to the concept of truth than Tarski's definitions tell us [Davidson, 1990, 285]. What is wrong is to conclude from this that Tarski has nothing to say about what the defined terms have in common, and nothing to say about the concept of truth.

How about the fourth objection, often called the "modal" objection [Raatikainen, 2008, 247]? According to the semantical definition, a sentence is true iff what it says is the case. Now, if "snow" had meant what "coal" currently means, "snow is white" would have been true iff what it said, namely, that coal is white, was the case. A deductive theory along Tarski's lines intended to apply to a language with that intuitive interpretation would be required to have '"snow is white" is true iff coal is white' as a theorem.[18] Tarski's account, taken as a whole, simply does not imply that truth is a property that "snow is white" would have iff snow were white, even if "snow" had meant what "coal" currently means.

What is true is that a definition along Tarski's lines for a language containing "snow is white" with the intuitive meaning that it has in English assigns the truth condition that snow be white to the sentence "snow is white" with no regard to the dependence of the accuracy of this truth condition on the intuitive interpretation imposed on the sentence by the conventions governing the object language and hence no sensitivity to changes in that intuitive interpretation. If the intuitive meaning of an expression of the object language changes, Tarski's definition needs to be reworked to capture the concept of truth for the the language so reinterpreted. This is a consequence of the fact that intuitive meaning is an "off table" matter to be kept in the mind of the users of the theory, or, to put it more in terms of the underlying philosophy of language, because the convention governing the use of the object language that maps "snow is white" to the thought that snow is what is something that is assumed rather than stated by the definition of truth for the object language.

That Tarski's theories are extensional is of course relevant here. Tarski's definitions, formulated within extensional theories, say nothing about what the truth conditions of sentences must or can be. Tarski's extensional theories are simply silent on the matter, and his definitions are intended simply to extend extensional theories. They make no claim to capture any supposed intension, in the contemporary sense of a function from worlds to extensions, of the truth-predicate.

The standard defenses of Tarski against the modal objection claim that his languages are individuated by the meanings of their expressions (e.g. Raatikainen, 2008). Is this the right response? Intuitionistic Formalism understands both object and metalanguage as governed by conventions that map sentences syntactically individuated to the thoughts that they express. A language identified by its set of sentences isn't individuated by the meanings of its expressions, but by these sentences. A language plus a set of conventions mapping expressions to concepts and sentences to thoughts is, in turn, of course individuated by the intuitive meanings of its expressions. I am thus inclined to split the difference with the standard response to the modal objection. What is right in the modal objection is that Tarski's definitions themselves say nothing about the dependence of truth conditions on intuitive meaning. The dependence of truth conditions on meaning, however, isn't absent from Tarski's overall account; it is present in the assumed conventions of the object language, conventions that are themselves not stated in the metatheory to which the truth definition is added. But just this fact blunts the force of the modal objection: Tarski in no sense seriously suggests that "snow is white" would be true iff snow is white no matter what "snow is white" meant. His view simply implies that this is a change in conventions requiring a corresponding change in extensional syntactic *cum* semantic theory. To that extent my reading implies something like the standard defense: the way the definition is laid out assumes that conventions fixing the meaning of the object language are held fixed. Where my view differs from the standard defense is that I do not attribute to Tarski the view that languages are individuated by the meaning of their expressions, since this is incompatible with his remarks about extending a language at [Tarski, 1983a, 164].

Moving on to the fifth objection, one common line of criticism of Tarski is that by adding definitions of semantic concepts to mathematical or logical theories he thereby makes the T-sentences (and instances of associated schemata for other semantic concepts) into mathematical or logical truths. Now what is clear is that given standard truth-conditional semantics and the theory-relative conception of definition, substitution in accord with a good definition doesn't preserve logical and mathematical truth and hence the fact that Tarski's definitions are added to mathematical theories is no reason to think that they on his view state mathematical truths. This was the argument of [Patterson, 2008b].[19] (Quick example: "an unmarried man is an unmarried man" is a logical truth, but "a bachelor is an unmarried man" isn't, even though the second is derived from the first by substitution in accord with a good definition—namely, itself.) But Intuitionistic Formalism muddies the waters significantly here since part of Tarski's view is that in a definition *definiens* and *definiendum* share their intuitive meaning.

One thought might be that a mathematical theory expresses only mathematical intuitive concepts before a Tarskian definition of truth for some object language is added to it; since the expression in terms of which "∈ *Tr*" is defined expressed a mathematical concept before the definition was added, adding the definition forces "∈ *Tr*" to express a mathematical concept. But this is no good, since it assumes that the account of intuitive meaning is compositional, which it isn't (cf. § 2.2.1). On the contrary, as I have argued in this chapter, the point of Convention T is to force an expression of a theory with only mathematical primitives to express a semantic concept. In accord with the theory-relative conception of definition combined with Tarski's account of the assignment of intuitive meanings to expressions constrained by theorems, once we add the Tarskian definition to a mathematical theory, what we have is a new theory that demands new interpretations for some of its expressions; in particular both "∈ *Tr*" and its *definiens* express the concept of truth. Hence on Tarski's view the T-sentences express semantic truths, not mathematical truths.

Now if Tarski had given some independent argument that semantic truths were actually mathematical truths it would be another matter, but since he didn't, we have nothing to go on but the obviousness of the point that semantic truths aren't mathematical truths, and the fact that Tarski had no need, in order to support any other position he held, to maintain that semantic truths are mathematical truths. Thus, I think, even taking Intuitionistic Formalism into account, the stronger view is that Tarski didn't hold that the T-sentences that follow from one of his definitions state logical or mathematical truths.

4.3.3 Reduction and physicalism

There is now a significant literature on whether Tarski's views are, as he appears to claim ([Tarski, 1983b, 406] where he also makes reference to another Vienna Circle doctrine, the "unity of science"), compatible with physicalism—whether he intended his definition to somehow support physicalism and what, according to him, physicalism amounted to. Passages other than the central one from ESS indicate reductive designs of some sort, e.g. the introduction to CTFL:

> In this construction I shall not make use of any semantical concept if I am not able previously to reduce it to other concepts [Tarski, 1983a, 153].

Yet, famously, Field [Field, 1972], seconded by others [Soames, 1984], argues that Tarski's definitions do not succeed in reducing semantic concepts to physicalistically acceptable ones. The problem lies again in Tarski's enumerative definition of satisfaction (and other sub-sentential semantic notions, if the language has non-logical lexical expressions other than predicates). A genuinely physicalist reduction of the notion of satifaction, for instance, would take the form

$$Sat(s, \vec{x}) \leftrightarrow \varphi(s, \vec{x})$$

where $\vec{x}$ is a sequence and φ is some genuinely physical predicate.[20] Field's preferred candidate, in good 1970s style, is the causal theory of reference [Field, 1972, 366–7],[21] but Tarski's "reduction" is obviously merely extensional [Field, 1972, 362].[22] As Field quips:

> By similar standards of reduction, one might prove that witchcraft is compatible with physicalism, as long as witches cast only a finite number of spells: for then 'cast a spell' can be defined without use of any of the terms of witchcraft theory, merely by listing all the witch-and-victim pairs [?, 369].

The first thing to note here, of course, is that by Tarski's own standards this was sufficient, since the theories were extensional.

However, in light of Field's criticism a substantial discussion has developed concerning the extent to which Tarski really intended any physicalistic reduction of semantic concepts or properties. Many authors discount Tarski's commitment to physicalism by noting that only in [Tarski, 1983b] is it suggested that his project aims to make semantics safe for physicalism [García-Carpintero, 1996, 117], [Frost-Arnold, 2004, 267–272] and that in this passage he is clearly trying to ingratiate himself to the audience at the Unity of Science Congress, Paris 1935.[23] As [Rojszczak, 1999, 123] points out, in the Polish version of the article, Tarski says, as in the German and Woodger's English translation of it, that if truth were axiomatized rather than defined there would be a threat the physicalism and the unity of science, but goes on to add, as those versions do not, that these are doctrines "propagated by a great many philosophers from the so-called Vienna Circle". This confirms the view that Tarski himself didn't have any particularly significant reductive goals for his semantics, but was simply ingratiating himself to the Vienna Circle.

On the view offered here, Tarski intended no reduction of any kind of the *concept* of truth: the intension or "content" of this concept is expressed in the irreducibly semantic "semantical definition" of [Tarski, 1983a, 155]. What he wanted to do was to construct deductive theories with only syntactic and logical (and hence mathematical, by arithmetization) primitives that, having the T-sentences for "$\in Tr$" for the object-language as theorems, constrained it to express this semantic content of the concept of truth. Tarski had no specific concerns at all about the content of the concept of truth itself. He took it to be familiar both from our everyday understanding of language and from philosophical works on truth. The issue was the conditions under which a term of a deductive theory that was itself conceptually unproblematic could express it consistently. Explicit definition, and hence ascent to a stronger metalanguage, were required for this last *desideratum* and the general thrust of metamathematics as he had understood that from well before the semantic period set the limits to the acceptable theories. The first paragraph of CTFL is clear on this point: though the content of the concept seems clear, "many investigations" of the problem of defining the content that begin from "apparently evident premises" break down in paradox. The issue wasn't getting straight about the concept, it was getting straight about the way in which it could consistently get expressed in a deductive theory.

4.3.4 Correspondence and deflationism

Interpreters are split on whether Tarski was a correspondence theorist or a deflationist. Popper, famously, was won over immediately by the page proofs of CTFL in Vienna and maintained thenceforth that Tarski had "rehabilitated the correspondence theory of absolute or objective truth" [Popper, 1963, 223]. More recent figures to maintain that Tarski's view is a correspondence theory include [Sher, 2004], [Sher, 1999], [Schantz, 1998], [Niiniluoto, 1999a], [Weingartner, 1999] and [Fernández-Moreno, 2001]. Most importantly, of course, Tarski himself says that the concept of truth in its "classical interpretation" involves the idea that "'true' signifies the same as 'corresponding with reality'" [Tarski, 1983a, 153], [Tarski, 1983b, 401].[24] Of course whether Tarski is a correspondence theorist depends on what a correspondence theory is and so in due course we will have to say a little about that.

There are two bad reasons and one better one to take Tarski to be a correspondence theorist. The bad reasons are the role of the T-sentences in Tarski's definitions and the role in these definitions of the relation of satisfaction that holds between open sentences and sequences. On the first count there is a fairly well established tradition of distinguishing "weak" from "strong" correspondence theories, where a "strong" theory posits some sort of structural relationship between sentences and the other relata of the correspondence relation, and a weak one merely insists that whether or not a sentence is true depends on whether what it says is the case [Woleński and Simons, 1989, 418] and, in different terminology, [Kirkham, 1992, 119], [Vision, 2004, 223]. Since the role of the T-sentences alone in Tarski's view gives one no reason to think that his view is a correspondence view in the strong sense—the T-sentences, after all, aren't even of the right form to ascribe a relation, much less one of structural similarity [Patterson, 2003]—the resultant readings insist on Tarski's being a weak correspondence theorist. (Kotarbiński himself criticizes strong correspondence theories just before introducing his version of Tarski's semantical definition at [Kotarbiński, 1966, 106–7].) This is correct as far as it goes, but only the common confusion [Jennings, 1983], [Vision, 2004, ch. 6], [Sluga, 1999, 40], [Sher, 1999, 136] on which deflationists somehow have an entirely "immanent" or "intralinguistic" conception of truth prevents the recognition that "correspondence" theories in the weak sense aren't even distinguished from deflationary theories.

As for satisfaction, it is often pressed into the service of correspondence readings of Tarski's definitions [Davidson, 2002b], [Woleński, 1989, 178], [Niiniluoto, 1999a], [Fernández-Moreno, 2001, 139]. The standard objection here is that by Def. 23 truth is satisfaction by all sequences and hence, by a given Tarskian definition, all true sentences correspond to the same thing, namely, the set of all sequences [Davidson, 1990, 303-4]. An objection heard less often, but equally important, is that for Tarski there is, strictly speaking, no relation of satisfaction in terms of which truth is defined, as we discussed in § 4.2.2. Add to this that n-place open sentences don't share a "form" with infinite sequences of objects in any interesting sense and we have an abundance of reasons that the appeal to satisfaction by correspondence readings of Tarski is fruitless.

The better reason to take Tarski to be a correspondence theorist is that he says that the idea behind his view is to capture the intuitions behind the "semantical definition" [Tarski, 1983a, 155] of truth and he is willing to equate these formulations with various correspondence formulations [Tarski, 1983a, 153], [Tarski, 1983b, 401]. The source of the previous poor construals of Tarski as a correspondence theorist is really again just the conflation of Tarski's "theory" with a particular definition, for with this conflation in place taking Tarski at his word that his view is a correspondence view mandates finding something in a particular definition or its consequences that amounts to a correspondence view. Properly distinguishing the content of the concept, the theorems of a deductive theory that force "$\in Tr$" to express it, and the definition that, added to the theory, results in its implication of these theorems immediately sorts the issue out. Neither the recursion on satisfaction nor the theoremhood of the T-sentences makes Tarski's "theory" a correspondence theory. Tarski is a correspondence theorist in the sense that he accepts the semantical definition and holds that the semantical definition defines a correspondence notion.

We turn next to the idea that Tarski was an early deflationist. There are many characterizations of deflationism in the literature. Without entering into the debate over how to define deflationism itself,[25] I will simply stick with the view I have maintained for some time now that the only thesis in the area clear enough to admit of meaningful dispute is the thesis that the T-sentences for a language define truth for that language [Patterson, 2002], [Patterson, 2008b].[26] Thus characterizing deflationism, Tarski's theory looks essentially deflationary as long as one stresses the idea that a definition of truth is the "logical product" [Tarski, 1983a, 187] of the T-sentences, as long as one sticks to common assumptions about what a definition is, and as long as one thinks that Tarski's definition is his theory. Quine [van Orman Quine, 1960, 24] attributes to Tarski the "classic development" of the insight that to say that "Brutus killed Caesar" is true is the same as to say that Brutus killed Caesar. [Leeds, 1978, 120–122] takes it that the T-sentences "axiomatize" the notion of truth and that Tarski's achievement was to discover as much. Horwich [Horwich, 1982, 191] distinguishes deflationists from "metaphysical realists" by their holding that the concept of truth is "exhausted" by Tarski's character-ization in terms of the T-sentences. Michael Devitt takes it as "generally agreed" that Tarski's "theory" is deflationary at [Devitt, 2001, 73].

Nevertheless, deflationists are often caused discomfort by the recur-sion on satisfaction in a standard Tarskian definition, so there are also many attempts to distinguish Tarski from deflationism in the literature—e.g. Horwich's later work [Horwich, 2003]. However, surely the overall emphasis on defining truth and related notions in terms of the instances of the T-schema looks consonant with what many contemporary defla-tionists champion. If only Tarski took his definitions and their conse-quences as conceptually analytic of truth, deflationary readings of Tarski would be in business. But, as argued here to this point, this isn't Tarski's view. The T-sentences constrain the interpretation of "$\in Tr$" down to be the content of the concept of truth, but they aren't themselves concep-tual analyses, and neither is the definition that brings them as theorems in its wake.

On the view offered here, the familiar clash of interpretations is entirely explicable. The requirement that the T-sentences be theorems if the concept of truth is to be expressed by "$\in Tr$" looks deflationary. Tarski's "semantical definition", coupled with Tarski's explicit claims that it is a correspondence theory, make Tarski look like a correspondence theorist. The muddle in the literature is unclarity over the relationship between Tarski's conception of truth and the theorems that make the "$\in Tr$" of a particular theory express it. Sorting out this muddle is one of the primary benefits of the interpretive work done in these first four chapters. The T-sentences don't directly state conceptual truths about truth; given Tarski's Intuitionistic Formalist account of intuitive meaning in terms of conventions, the theoremhood of the T-sentences constrains "$\in Tr$" to express the content of the concept of truth as explicitly stated in the semantical definition. Tarski was a correspondence theorist about the concept of truth, granting the claim that the semantical defini-tion is equivalent to various correspondence formulations. However, Tarski isn't a correspondence theorist in the sense that he posits some sort of isomorphism or picturing-relation between true sentences and facts or something of the sort. The upshot of his Intuitionistic Formalist views is that a theory with the T-sentences as theorems is sufficient to express that concept, and this theory looks like a deflationary theory. But Tarski doesn't hold that the theory defines the concept, merely that it constrains "∈ *Tr*" to express it. Hence his view lacks most of the consequences of deflationary views, e.g. the implication that strictly speaking there are different concepts of truth for different languages.

So is Tarski a correspondence theorist, at least of the content of the concept of truth that his various definitions aim to capture relative to their object languages? This depends on whether "*x* is true iff *x* says that p, and p" is itself, as Tarski said it was, equivalent to correspondence conceptions of truth. It certainly isn't equivalent to a "strong" such conception, since it makes appeal to no notion of similarity in form. It does, as "weak" correspondence requires, allow that truth depends on extralinguistic fact, but as noted above this doesn't distinguish it from deflationary theories. For my money [Patterson, 2008b], [Patterson, 2005], the semantical definition is deflationary since it assumes some meaning-theoretic notion and explains truth in terms of it, while only "strong" correspondence theories really deserve to be called "correspondence theories". In any case, the point of this sub-section stands: Tarski takes the semantical definition to be a correspondence conception of truth, and aims to capture it in deductive theories that have the T-sentences for their object languages as consequences. As long as we don't conflate Tarski's particular definitions with his "theory" and as long as we recognize that a particular definition isn't supposed to state anything conceptually analytic about truth, and that the T-sentences act as constraints that, in conjunction with Tarski's Intuitionistic Formalist conception of intuitive meaning, hedge "∈ *Tr*" about with sufficient restrictions to force it to express the content of the concept of truth as construed in the semantical definition rather than directly analyzing this content, we can appreciate the overall structure of concept, definition and theory that gives rise to the varying claims about truth that have hitherto caused interpreters so much difficulty.

5
Indefinability and Inconsistency

Our discussion of Tarski's Intuitionistic Formalist treatment of truth in §§ 2–4 of CTFL concerned material that seems to have been developed by 1931 ([Hodges, 2008, 120-6]) and merely polished for the 1931 submission of the Polish original of the work. (Cf. Tarski's own "Historical Note" where the addition of the Gödelian treatment of indefinability is described as, of all the results in the work "the only one subsequently added to the otherwise already finished investigation" [Tarski, 1983a, 278].) We have thus been able to discuss the material of §§ 2–4 with little concern for our overall story about Tarski's development from the late 1920s up to the Paris Unity of Science conference in 1935. Now that we turn to Tarski's remarks on the conditions under which truth cannot be defined we will be unable to ignore the issues raised by Tarski's evolving views.

In particular, three different views on the impossibility of constructing an intuitively adequate definition of truth for "expressively rich" languages can be found in CTFL as we have it: a pre-Gödelian argument that an intuitively adequate truth-definition is impossible based on considerations from the theory of semantical categories, a basically Gödelian argument that appears in the "proof-sketch" accompanying Theorem I of § 5—though one still conceived in terms of Tarski's original Intuitionistic Formalist metamathematics—and a reassessment of the § 5 results set out in the postscript to the German translation of 1935, which begins to pave the way for an abandonment of Intuitionistic Formalism that begins in earnest with the treatment of consequence in 1935. Even this categorization is too simple: the argument for Theorem I is a mess that combines aspects of Gödel's presentation with bits of Tarski's interest in formal definability, and the text of the passage itself is a patchwork attempt to paste the later result into a work that was already written to lead up to the earlier one.

5.1   Indefinability

5.1.1   Indefinability before 1931

What Tarski says about the theory of semantical categories in § 4 already implies a kind of indefinability theorem: no language the order of the variables of which has no finite upper bound can receive a definition of truth along the lines so far developed, because the satisfaction predicate must be of higher order than any variable in the language. In § 5 Tarski introduces "the general theory of classes", which has variables of this description and atomic open sentences, Tarski's intuitive reading of which is "the class X (of $n+1$th order) has as an element the object Y" or "The object Y has the property X"' [Tarski, 1983a, 242–3]. Tarski states the argument forthrightly after introducing the language and the axioms for the theory:

> When we try to define the concept of satisfaction in connexion with the present language, we encounter difficulties which we cannot overcome. In the face of the infinite diversity of semantical categories which are represented in the language the use of the method of many-rowed sequences is excluded from the beginning, just as it was in the case of the logic of many-termed relations. But the situation here is still worse, because the method of the semantical unification of the variables also fails us. As we learnt in § 4, the unifying category cannot be of lower order than any one of the variables in the language studied. Sequences whose terms belong to this category, and still more the relation of satisfaction, which holds between such sequences and the corresponding sentential functions, must thus be of higher order than all those variables. In the language with which we are now dealing variables of arbitrarily high (finite) order occur: consequently in applying the method of unification it would be necessary to operate with expressions of 'infinite order'. Yet neither the metalanguage which forms the basis of the present investigations, nor any other of the existing languages, contains such expressions. It is in fact not at all clear what intuitive meaning could be given to such expressions [Tarski, 1983a, 243–4] (altered).

The 1931 project screeches to a halt at the limits of intuitive meaning.[1] This is bound by the theory of semantical categories, and this simply leaves no room for the idea of an expression of "infinite" order. Tarski briefly considers doing things in a piecemeal fashion [Tarski, 1983a, 244–5]: since each particular sentential function has free variables of some fixed finite order, one might hope for some sort of recursion on the order of open sentences. However, this hope is dashed by the point that open sentences of a certain order may contain bound variables of arbitrarily high order [Tarski, 1983a, 245]. Tarski concludes on a low note: "Therefore we can say—considering the failure of previous attempts—that at present we can construct no correct and intuitively adequate (*sachlich zutreffende, zgodnej z intuicją*) definition of truth for the language under investigation" [Tarski, 1983a, 245–6].

My hypothesis is that when the work originally went to press this was its final sentence. The published work, however, does not end there. Tarski notes that Theorem I and the sketch of its proof were added to the work "after it had already gone to press". Given that other concerns in the rest of § 5—the ω-rule, the possibility of developing an axiomatic theory of truth—are also foreign to the concerns of the earlier sections, I think the best guess is that everything in § 5 after the top of page 246 was added to what was supposed to have been a finished work and that the second, longer proof sketch for Theorem I at 249–50 was added even later. As we will see, the results of this piecemeal composition show.

As he himself notes, Tarski was moved to extend the work by the appearance of the 1930 report announcing Gödel's 1931 article presenting the incompleteness theorem. Though Tarski himself had independently recognized the basic possibility of arithmetization and mentioned it at [Tarski, 1983a, 184], Gödel's report prompted Tarski to extend his work along Gödel's lines and to amend the work in press to get the results into print.

The first thing Tarski does in the added material is to ask a question in a general form that is clearly raised by his original downbeat conclusion, but that the material up to that point in the work put him in no position at all to address: "*whether on the basis of the metatheory of the language we are considering the construction of an adequate (richtige, trafnej) definition of truth in the sense of Convention T is in principle possible*" [Tarski, 1983a, 246] (corrected). Before encountering Gödel's techniques, all Tarski was in a position to say was that his techniques could be applied within the bounds set by the theory of semantical categories only to languages of finite order. If nothing else were possible, one would like to know that, but nothing Tarski had before late 1930 put him in a position to address the question.

This sets the stage also for Tarski's early adaptation of Gödel's results. Tarski's question all along had been whether semantic concepts could be formally expressed by definitionally eliminable terms. This means that when Gödel first allowed him to extend his investigations the techniques were viewed entirely in light of the question Tarski had already been asking: can the intuitive concept of truth be expressed by any *explicitly defined* expression of the metalanguage under the conditions set out in § 5? In particular, in first appropriating Gödel's work, Tarski views it and the result that now bears Tarski's name not as it is viewed today, as a semantic result about whether an expression can semantically define the *set* of true sentences, but as a formal result about whether any expression of the metalanguage can be intuitively adequate to express the content of the concept of truth through the T-sentences for all sentences of the object language being theorems. It took Tarski's thinking a number of years to emerge from the 1929–1930 project and see the results in a fully contemporary semantic light, and along the way the early conception of meaning, to which the semantics for which Tarski is now remembered was a mere auxiliary, had to be superseded by this same semantics. The version of the results present in the work as we know it is, by contrast, thoroughly colored by his conception of it as a contribution to his project of formal definition and, ultimately, the expression in a deductive theory of intuitive semantic concepts as he understood them.[2]

5.1.2 Theorem I: textual issues

The usual view of Theorem I and the proof-sketches that follow it is that therein Tarski states an early form of "Tarski's Theorem", according to which no one-place open sentence in the vocabulary of a theory that extends a weak system of arithmetic can have as its extension exactly the set of Gödel numbers of the theory's truths. Sometimes it is recognized that the text is outdated to some extent, e.g. in its formulation in STT, but by and large this is the extent of the appreciation of the passage.

"Tarski's Theorem" as we now know it is a theorem about semantic definability. Consider a weak system of first-order arithmetic like Robinson's Q and any theory T that extends it. Gödel code the language of T, representing the Gödel code of an expression e as $\langle e \rangle$, and consider the set of Gödel codes of sentences s such that $T \models s$. Suppose that ψ semantically defines this set relative to T, that is, that

$$\text{(REP)} \quad T \models s \text{ if and only if } T \models \psi \langle s \rangle$$

Now, by the diagonal lemma for extensions of Q, for every open sentence Fx there is a sentence s such that $T \vdash s \leftrightarrow F \langle s \rangle$. So, with respect to ψ there exists a sentence s_0 such that

$$(\text{DIAG}) \quad T \vdash s_0 \leftrightarrow \neg \psi \langle s_0 \rangle$$

Now, suppose that $T \models s_0$. Then, by the soundness of $\vdash$ relative to $\models$ and closure of the latter under *modus ponens*, if $T \models s_0$ and $T \vdash s_0 \leftrightarrow \neg \psi \langle s_0 \rangle$ then $T \models \neg \psi \langle s_0 \rangle$. Thus:

$$T \models s_0 \text{ if and only if } T \models \neg \psi \langle s_0 \rangle$$

However, by (REP):

$$T \models s_0 \text{ if and only if } T \models \psi \langle s_0 \rangle$$

Therefore:

$$T \models \psi \langle s_0 \rangle \text{ if and only if } T \models \neg \psi \langle s_0 \rangle$$

Hence, if (DIAG) and (REP) are true, T is semantically inconsistent. So if T is a semantically consistent extension of Q, (REP) is not true. There is thus no ψ in the language of T that semantically defines the set of sentences s such that $T \models s$. If we further assume that T is true (and that $\models$ preserves truth), we have Tarski's indefinability theorem: no theory that extends Q can semantically define the set of its own truths. This argument makes no use of any assumption that ψ has an explicit definition that, when added to T, implies the T-sentences $\psi \langle s \rangle \leftrightarrow s$. The assumption for reductio is, rather, (REP).

Though basically this argument is often attributed to the passage, there are serious questions to be raised as to whether it is actually to be found there. In [Patterson, 2008a] I contrasted two readings of the passage, what I called a 'semantic' one which at least finds some analogue of the contemporary argument in the passage, a version of which can be found in [Patterson, 2006], and a 'syntactic' one which is close to that offered by [Gómez-Torrente, 2004] and which finds nothing like the contemporary argument in the passage. ([Vaught, 1986, 871–2] gives a similar reconstruction much more briefly, claiming in addition, as Gómez-Torrente does as well, that the notion of truth really plays no role in the argument). At the time, as stated then, I was simply unsure how to make sense of the passage, since neither reading seemed able to to account for all of its features. I have since become somewhat more sympathetic to the syntactic reading—not enough so to think that it can declare the passage as we have it coherent, but enough to think that the syntactic reading's argument must be what Tarski had in mind and that other features of the passage are probably just to be chalked up to hasty composition and Tarski's perhaps imperfect grasp on the issues at the time.

The result to be established is stated thus:

Theorem I (α) *In whatever way the symbol 'Tr', denoting a class of expressions, is defined in the metatheory, it will be possible to derive from it the negation of one of the sentences which were described in the condition (α) of the Convention T;*

(β) *assuming that the class of all provable sentences of the metatheory is consistent, it is impossible to construct an adequate definition of truth in the sense of Convention T on the basis of the metatheory* [Tarski, 1983a, 247].

Notice that this assumes that "*Tr*" has an explicit definition. The main argument then runs as follows. Tarski points out that the natural numbers can be defined in the general theory of classes and that arithmetic can be expressed there; he comments on the identification of numbers with classes of equinumerous classes and then reminds the reader of the possibility of enumerating the sentences of the object language and thereby arithmetizing the metalanguage's formal syntax [Tarski, 1983a, 249]. Then comes the argument:

Let us suppose that we have defined the class *Tr* of sentences in the metalanguage. There would then correspond to this class a class of natural numbers which is defined exclusively in terms of arithmetic. Consider the expression '$\bigcup_1^3 (\iota_n.\phi_n) \bar{\in} Tr$'. This is a sentential function of the metalanguage which contains '*n*' as the only free variable. From the previous remarks it follows that with this function we can correlate another function which is equivalent to it for any value of '*n*', but which is expressed completely in terms of arithmetic. We shall write this new function in the schematic form '$\psi(n)$'. This we have:

(1) *for any n,* $\bigcup_1^3 (\iota_n.\phi_n) \bar{\in} Tr$ *if and only if* $\psi(n)$.

Since the language of the general theory of classes suffices for the foundation of the arithmetic of the natural numbers, we can assume that '$\psi(n)$' is one of the functions of this language. The function '$\psi(n)$' will thus be a term of the sequence ϕ, e.g. the term with the index *k*, '$\psi(n)$' $= \phi_k$. If we substitute '*k*' for '*n*' in the sentence (1) we obtain:

(2) $\bigcup_1^3 (\iota_k.\phi_k) \bar{\in} Tr$ *if and only if* $\psi(k)$.

The symbol '$\bigcup_1^3 (\iota_k.\phi_k)$' denotes, of course, a sentence of the language under investigation. By applying to this sentence condition (α) of the Convention T we obtain a sentence of the form '*x* $\in$ *Tr if and only if p'*, where '*x*' is to be replaced by a structural-descriptive or any other individual name of the statement $\bigcup_1^3 (\iota_k.\phi_k)$, but '*p*' by this statement itself for by any statement which is equivalent to it. In particular we can substitute '$\bigcup_1^3 (\iota_k.\phi_k)$' for '*x*' and for '*p*'—in view of the meaning of the symbol 'ι_k'—the statement 'there is an *n* such that $n = k$ and ψ_n' or, simply 'ψ_k'. In this way we obtain the following formulation:

$$(3) \bigcup_1^3 (\iota_k.\phi_k) \in Tr \text{ if and only if } \psi_k.$$

The sentences (2) and (3) stand in palpable contradiction to one another; the sentence (2) is in fact directly equivalent to the negation of (3). In this way we have proved the first part of the theorem. We have proved that among the consequences of the definition of the symbol '*Tr*' the negation of one of the sentences mentioned in the condition (α) of the Convention T must appear. From this the second part of the theorem immediately follows [Tarski, 1983a, 250–1].

The syntactic reading takes the passage to be a relatively straightforward argument for the theorem as stated, though it has to posit some expository hiccups. It represents the passage as arguing something like this:

We can enumerate the expressions of the object language in the metalanguage. Now consider the metalanguage open sentence 'the nth sentence is not *F*', for arbitrary *F*. arithmetization tells us that this is equivalent in the metalanguage to some arithmetic open sentence $\psi(n)$. Hence we have 'The nth sentence is not $F \leftrightarrow \psi(n)$'. Also by arithmetization, $\psi(n)$ is in our enumeration; let us suppose it is the kth expression. Instantiating *n* to *k*, we have (2) 'the kth sentence is not $F \leftrightarrow \psi(k)$'. Now, suppose that *F* is the expression we've introduced by a definition that satisfies Convention T. Then the T-sentence (3) 'the kth sentence is $F \leftrightarrow \psi(k)$' follows from the definition of truth and the rest of the metatheory. But (2) is the negation of the T-sentence (3), and so our theory is rendered inconsistent by the assumed intuitively adequate definition of '*F*'.

This argument makes sense, shows roughly what the explicit statement of the theorem says is to be shown, and runs as it should run if the point is an application of a general point about arithmetization being sufficient for the metatheory to imply the negation of a T-sentence. However, there is a great deal about the passage that doesn't sit well with the reading.

Tarski begins the argument by assuming that the "we have defined the class *Tr* of sentences of the metalanguage". If the point were simply the syntactic one about diagonalization and the T-sentences, there would be no need for this: the point need simply be that *for every* predicate F there is a sentence s such that $s \leftrightarrow F\langle s \rangle$ is a theorem. Gómez-Torrente's "anachronistic reconstruction" represents the argument as doing exactly this [Gómez-Torrente, 2004, 31]—Gómez-Torrente simply doesn't address the fact that Tarski not only assumes that "*Tr*" is at issue, but that the assumption is that the *class Tr* of sentences has been defined in the metalanguage (the German and Polish versions agree on this).

More worryingly, Tarski's argument for (α) of Th. 1 is done with displayed thesis (2) (as Gómez-Torrente notes [Gómez-Torrente, 2004, 34]). But he goes on to arrive at (3) and *then* says "in this way we have proved the first part of the theorem". But what, then, is the argument for (3) there to establish? Likewise, he also says that "from this the second part of the theorem immediately follows". But the second part of the theorem, (β), says that if we do add a definition that implies the T-sentences to the theory, the theory will be rendered inconsistent. But this doesn't just immediately follow, given that (3) is already on board: it has already been shown—if, that is, the text that precedes (3) is an argument that if we add an intuitively adequate definition of truth the T-sentence will be a theorem.

Further problems arise when we try to address these issues. Suppose we hold that (3) and the sentences leading up to it are there merely to establish that (2) *is* the negation of a T-sentence. This would make sense of Tarski's saying that (β) remains to be shown, but follows immediately: (3) hasn't been claimed to follow from the metatheory yet; it was just mentioned by way of establishing that (2) is the negation of a T-sentence. However, it makes less sense of what status (1) through (3) are supposed to have. Tarski says in a note on 249 that "instead of saying that a given sentence is provable in the metatheory, we shall simply assert the sentence itself" in the proof-sketch. This would seem to mean that we're to take the claim to be that (1) through (3) follow from the metatheory. This makes sense for (1) and (2), but it sticks us with the view that (3) is claimed to follow from the metatheory as are (1) and (2). More generally, there is no assumption in the proof that "*Tr*" has a definition adequate according to Convention T; the theorem is supposed to concern a predicate formally defined "in whatever way", and, again, at 251 we're told that it remains (easily) to be shown that *if* we add a definition adequate according to Convention T the metatheory is rendered inconsistent. Hence the suggested way of making sense of (3)'s role and the claim that (β) is left to be shown isn't compatible with [Tarski, 1983a, 249 note 1].

Finally, there is a further issue of historical accuracy here concerning in exactly how diagonalization plays a role in the argument. Both the readings in [Gómez-Torrente, 2004] and [Vaught, 1986, 871] attribute an argument involving a a general form of the diagonal lemma to Tarski's text. Now Tarski does speak of "applying the diagonal procedure from the theory of sets" in the shorter proof-sketch at [Tarski, 1983a, 248], but the argument of the longer proof-sketch fails to involve any general claim amounting to (DIAG). Tarski cites [Fraenkel, 1928, 48], which discusses, as Tarski says, the use of diagonal constructions in set theory, but there is no statement of the diagonal lemma there—not surprisingly, since Fraenkel is concerned with set theory rather than proof-theory. [Vaught, 1986, 871] notes the similarity to [Gödel, 1967b], but Gödel, like Tarski, concerns himself only with particular instances of (DIAG). Gödel himself later attributed the general diagonal lemma neither to himself nor to Tarski, but to Carnap's *Logical Syntax* in a footnote added later to the write-up of his 1934 Princeton lectures [Gödel, 1986, 363 note 23]. Thus, no reading that attributes some appeal to a general diagonal lemma to the passage can be accurate.[3]

These are the problems faced by the syntactic reading. As mentioned, I've come around to thinking that the syntactic reading must capture what Tarski had in mind. Before saying why, let us look briefly at how the semantic reading would interpret the passage. The idea is that the argument looks like this:

> Suppose that Tr is the set of truths of the object language. Now by diagonalization there is a sentence s such that $s \leftrightarrow \neg Tr\langle s \rangle$ is a theorem of our theory. (So far, this is the reading of (1) and (2).) Now, since Tr is the set of truths of the object language, if s is a theorem, so is $Tr\langle s \rangle$ and vice versa, since we're assuming that our theory is true and assuming that Tr defines the set of truths. So $s \leftrightarrow Tr\langle s \rangle$ is also a theorem. (That's (3).) Hence, if in fact Tr semantically defines the set of truths, our theory is inconsistent. So if our theory is consistent, nothing in it semantically defines the set of truths.

The merit of this reading is that it makes the progression from (1) to (3) on page 250 a sensible argument for a conclusion that Tarski could have been proving in 1931 in view of Gödel's work: it makes use of the assumption that the set Tr of truths is defined semantically, it gives (3) an obvious job to do, it makes sense of Tarski's claim in the note to 249 that he'll state theses meaning by them that they are theorems of the metatheory, and it makes sense of the passage between (2) and (3) as an argument that (3) is a theorem without importing the assumption that it is so because "*Tr*" has a formal definition that is intuitively adequate according to Convention T, thereby leaving the point about Convention T yet to be shown, just as it is claimed to be on 251. (Why is Convention T even mentioned then? Simply to remind the reader of the T-sentences: notice that Tarski just says "apply condition (α) of Convention T" not something like "apply the assumption that *Tr* has an explicit definition that is intuitively adequate according to Convention T" and that he similarly refers to the T-sentences this way in the statements of Theorems II and III.)

The weakness of this reading is that the argument it attributes to the passage does more than to prove (α) and set up for an immediate proof of (β): it proves something that Theorem I says nothing about, namely that no one-place open sentence of the metatheory can semantically define the set of truths. Theorem I, as stated, simply isn't about this.

The reason that I have come to favor the syntactic reading is that I have come around to the story I'm telling in this book: it was not until well after the 1931 publication of the work that Tarski really came to see semantics as he had introduced it as his central concern. In 1931 Tarski was concerned with giving formal definitions of semantic terms that were intuitively adequate by his standards. To show that no such definition is possible in the case at hand in § 5, Tarski only needs the syntactic theorem, and since his semantics was merely a tool for crafting formal definitions of semantic notions, he simply wouldn't have seen semantic definition of the set of truths as the central issue.[4]

So the reading of the passage I would now favor is that Tarski at least wants the "syntactic" result, since that is the result stated by Th. I, and since that is what, on my general story here, he would have been interested in in 1931–3. The text we have on 250, however, is far from a clear attempt to do that. Notice also that in 1931–3 Tarski is not yet working comfortably with a distinction between syntactic and semantic consequence. Since it was only several years later that he was to formulate the latter notion and clearly to distinguish it from the former, it isn't terribly surprising to find him failing to be clear about it in the passage at hand.

In light of this the apparent shift from a syntactic theorem to a semantic proof seems explicable, since the argument on 250 is clearly lifted from § 1 of [Gödel, 1967b]. Gödel's informal presentation of the result that he goes on to prove in detail runs as follows: consider the class K of numbers n such that it isn't provable in P (A version of STT plus Peano Arithmetic) that the nth formula in some enumeration of formulas applies to them. Some formula of P, say the qth, defines K; call it $R(q)$, and let $[R(q), n]$ be the formula in which the free variable in $R(q)$ is replaced by a numeral for n. Now suppose that $[R(q), q]$ is provable in P. Then $q \in K$. But by hypothesis, if $q \in K$ then $[R(q), q]$ is not provable in P. Contradiction. So Suppose that $\neg[R(q), q]$ is provable in P. Then, by hypothesis $q \notin K$ and thus $\neg[R(q), q]$ is provable after all. Contradiction. So neither $[R(q), q]$ nor $\neg[R(q), q]$ is provable in P. Gödel goes on to note that since $[R(q), q]$ is not provable, it is true [Gödel, 1967b, 598].

This argument involves the back-and-forth between a formula and the set it semantically defines that is characteristic of the modern take on the results, and that is attributed to page 250 of Tarski's work with some justice by the semantic reading. Since Tarski by his own admission [Tarski, 1983a, 275] was spurred to add Theorem I and the proof sketches by [Gödel, 1967b], we can imagine him working by noting that, as he says, "the results obtained for the system P can easily be carried over to the present discussion" [Tarski, 1983a, 248]. With the goal of showing that no intuitively adequate definition of truth could be supplied for the language of the general theory of classes, Tarski needed to recast Gödel's argument to concern truth rather than provability and to involve the derivation of the negation of a T-sentence; what results is something that began life in Gödel's hands as concerning semantic definition of a set but becomes, in Tarski's hands a garbled combination of that with a point about formal definition. If we remember that these things were new then, the uneven character of the passage—which, on balance, seems to be a proof of the semantic result inserted into a discussion of the syntactic result—is understandable: Tarski was trying to carry over an argument for the semantic result and put it to syntactic purposes and didn't do a particularly thorough job of it. He sees that the first part of Gödel's argument can easily be recast as a proof that the negation of a T-sentence is provable with suitable adjustments, but leaves in from Gödel's presentation the assumption of the definition of a set and some argument that an inconsistency results given the definition of the set and the result of the first part of the proof.

Indeed, a further aspect of 247–251 might explain why an apparent proof of a more Gödelian result appears as the proof of Theorem I. After introducing the theorem, on pages 247–8 Tarski gives a *very* general sketch of the basic application of diagonalization to produce the negation of a T-sentence. He then, rather oddly, says at the top of 249 "We shall sketch the proof a little more exactly" and starts over with the proof-sketch we have been discussing.[5] This stuttering presentation invites the following hypothesis: Tarski first wrote the additions to CTFL that follow page 246 using only the first, short proof sketch, very soon after learning of the basic thrust of Gödel's results. Perhaps he even sent the extension of § 5 to the printer in that form. After composing this material Tarski continued to study Gödel's results, trying out various things himself and exploring various lines; since Gödel's discussion really is a semantic one, and since Tarski was prepared to understand it that way, such considerations would have been to varying extents along Gödel's semantic lines. At some point it occurred to Tarski that it would be good to have something a little more concrete as a proof-sketch than what we now have on page 248, so he inserted the second, longer proof-sketch, which we can think of as perhaps something taken from working notes. Somehow he wasn't careful to make the later addition match the surrounding text.

Whatever the reason, this would explain the way in which the material on 250 is more or less an independent proof of a result different from, but closely related to, Theorem I, and its poor fit with what he says he has done and has left to be done at the top of 251. In favor of this reading, note that if one skips from the last sentence of 248 to the sentence that begins on the last line of 250, one produces a discussion of Theorem I and its "proof" that makes perfectly good sense: on page 248 we're told that diagonalization produces the negation of a T-sentence, thereby establishing (α), and then 251 makes the point that (β) follows easily, since adding an intuitively adequate definition will bring the corresponding T-sentence in its wake. What lies between those two points is a passage that stands alone as the proof of a semantic indefinability theorem much more like the contemporary take on "Tarski's Theorem".

This is speculative, of course. My best sources tell me that they know of no extant proof sheets or manuscripts that would answer questions about the order of composition of the passages. But the story makes enough sense of the problematic features of the passage from 247–251 that I am inclined to accept it until evidence against it comes in.

5.1.3 Theorem I and Intuitionistic Formalism

As I said above, I have come to think that the theorem according to the syntactic reading is what Tarski intended to prove. This is because it matches much more closely the animating concerns of CTFL than does the semantic result. By what I have posited is the original end of the work—the top of page 246 in *Logic, Semantics, Metamathematics*—Tarski is in the following position. His goal is explicitly to define "$\in Tr$" so that the T-sentences involving it are theorems; the requirement is that the definition should be logically in order ("formally correct") and that when added to the background theory the T-sentences should become theorems ("intuitively adequate"). Given the embedding of the construction in STT and Tarski's commitment to this embedding as consonant with the requirements of the theory of semantical categories and, thereby, the basic structure of intuitive meaning, it turns out that only for languages of finite order can satisfaction be defined on the way to defining truth as satisfaction by all sequences.

Tarski's view requires that the T-sentences be theorems; the requirement that "is true" explicitly be defined, and moreover so defined against a syntactic (ultimately, arithmetic) background theory comes not from the desire not to express the intuitive conception of truth, but to express it consistently—that is, to ward off semantic paradox. The theory of semantic categories as Tarski applies it—only as applying to expressions with "independent meaning" and, for the purposes of the work, only to variables and predicates—doesn't pass any particular verdict on the T-sentences themselves to the effect that the metalanguage can contain them only if it is of higher order than the object language. So what Tarski has with the passage to a language of "infinite" order is simply that if the T-sentences are to be introduced, it can't be by way of any relation of satisfaction, which according to the theory of semantical categories does need to be of higher order than all the variables of the object language. He is thus left with a question: is there any way to get the T-sentences consistently into the metalanguage when the object language is of infinite order? Is there, in particular, any way to introduce a definition that implies the T-sentences consistently in such a case?

Theorem I and the argument for it function to show that it isn't just some weakness of the approach of defining truth by recursion on satisfaction that leads the project to grind to a halt at infinite order; diagonalization shows that the T-sentences cannot be theorems for *any* arithmetically definable predicate in full STT. In the abstract, this leaves a choice of views: either truth isn't arithmetically definable, or the T-sentences aren't theorems. The point of interest to us is that Tarski accepts the first rather than the second view, thereby setting himself at odds with nearly all contemporary logical treatments of semantic paradox. As Tarski already begins to recognize more clearly in the Postscript, the real upshot of Gödel's results is that the metatheory can have a predicate that semantically defines the set of truths of the object language only if the metalanguage can semantically define more sets than the object language can, since the result is that no predicate of the object language can do so. Now, suppose we're interested in the case in which the metatheory can't define more sets, say because we're interested in the case in which a language is, in Tarski's 1944 terms, "semantically closed"—perhaps because we think that natural languages are semantically closed and we want to model them in this respect. Here we have the two options that confront Tarski in interpreting his result in 1931–3: we can say that there is no set of truths, or we can say that there is one, but that that at least some of the T-sentences are untrue. In the contemporary literature, the second option is favored. On the treatment in [Kripke, 1975], for instance, the object language contains a predicate that semantically defines the set of its own truths, but some of the T-sentences in the object language are untrue (they're untrue for ungrounded sentences, which include paradoxical sentences). Of course, in Kripke's metalanguage the T-sentences are true, but the metalanguage can define more sets than the object language can, in particular the set of untruths. Most contemporary work follows Kripke in giving up some T-sentences in order to get a predicate that semantically defines the set of truths into the object language.

Tarski takes the other route. The whole point of the exercise, for Tarski, is to express the concept of truth in accord with the semantical definition in a deductive theory. This requires that the T-sentences be theorems. If we are to proceed with this project in the face of Theorem I, the only option is to give up on the arithmetic definability of truth. This is just what Tarski does. He expresses himself as follows:

> The result reached in Th. I seems perhaps at first sight uncommonly paradoxical. This impression will doubtless be weakened as soon as we recall the fundamental distinction between the content of the concept to be defined and the nature of those concepts which are at our disposal for the construction of the definition ...
>
> Only thanks to the special methods of construction which we developed in §§ 3 and 4 have we succeeded in carrying out the required reduction of semantical concepts, and then only for a specified group of languages which are poor in grammatical forms and have a restricted equipment of semantical categories ... The analysis of the proof of Th. I sketched above shows that this circumstance is not an accidental one. Under certain general assumptions, it proves to be impossible to construct an adequate (*richtige, trafnej*) definition of truth if only such categories are used which appear in the language under consideration [Tarski, 1983a, 251–4] (corrected).

The result is apt to strike us today as unremarkable. Why, for Tarski at the time, was it "uncommonly paradoxical"? Precisely because he was perfectly confident that the intuitive concept of truth could be applied to meaningful sentences, and the language of the general theory of classes consists of meaningful sentences. So what is "paradoxical" about the proof is its apparent clash with the semantical definition of truth, a conception that Tarski took to be obviously correct. The rest of the passage explains away *that* appearance of paradox: Theorem I doesn't impugn our intuitive conception of truth, it impugns the attempt to express it in a theory that contains only "structural descriptive" terms, that is, it impugns the assumption that an expression that expresses the content of the concept of truth (as expressed in the semantical definition) can be eliminatively defined in a mathematical background theory when the object language can define all the sets that the metalanguage can. The validity of the semantical definition of truth and the requirement of the theoremhood of the T-sentences goes unquestioned in all of this. Tarski continues:

> An interpretation of Th. I which went beyond the limits given would not be justified. In particular it would be incorrect to infer the impossibility of operating consistently and in agreement with intuition (*mit der Intuition übereinstimmenden, zgodnego z intuicjką*) with semantical concepts and especially with the concept of truth [Tarski, 1983a, 255].

Note the recurrence of the phrase that ended the original manuscript [Tarski, 1983a, 246], something obscured by Woodger who, following Blaustien, varies the translation of "*zgodnego z intuicjką*" between the two passages.

5.1.4   Axiomatic semantics

Explicit syntactic (arithmetic) definition being out of the question, Tarski notes that that it is possible to define truth for sublanguages of finite order and then contemplates a move to axiomatic truth-theory for the language of the general theory of classes as a whole:

> Theorem III *If the class of all provable sentences of the metatheory is consistent and if we add to the metatheory the symbol 'Tr' as a new primitive sign, and all theorems which are described in conditions (α) and (β) of the Convention T as new axioms, then the class of provable sentences in the metatheory enlarged in this way will also be consistent* [Tarski, 1983a, 256].

Since the metatheory is only contradictory if some finite subset of it is, and since any finite set of T-sentences can be made the consequences of an explicit definition, the sentences these T-sentences concern being of only finite order, adding denumerably many T-sentences will not render the metatheory inconsistent.

However, as Tarski goes on to note, the result is by his standards logically and conceptually unsatisfactory. General theses whose provability

Tarski demands aren't available—one cannot, for instance, prove that no sentence of the object language is both true and untrue, though one can prove of each that it isn't both true and untrue.[6] Simply adding what we want in the form of additional axioms won't address the real issue, which for Tarski is that the axiom system isn't provably categorical in the sense of § 2 of DC:

> It seems natural to require that the axioms of the theory of truth, together with the original axioms of the metatheory, should constitute a categorical system. It can be shown that this postulate coincides in the present case with another postulate, according to which the axiom system of the theory of truth should unambiguously determine the extension of the symbol '*Tr*' which occurs in it, and in the following sense: if we introduce into the metatheory, alongside this symbol, another primitive sign, e.g. the symbol '*Tr'*' and set up analogous axioms for it, then the statement '*Tr* = *Tr'*' must be provable. But this postulate cannot be satisfied. For it is not difficult to prove that in the contrary case the concept of truth could be defined exclusively by means of terms belonging to the morphology of language, which would be in palpable contradiction to Th. I [Tarski, 1983a, 258].

Since the theory that results from adding the T-sentences and any other sentences we please is either inconsistent or not provably categorical, we can only incompletely express the concept of truth axiomatically. So in moving to axiomatic truth theory, we give up all of Tarski's logical and expressive goals: our theory isn't finitely axiomatizable, we don't express the content of the concept of truth in a theory that has unproblematically mathematical primitive expressions, and even at that cost we don't completely express it. Tarski is thereby unenthusiastic about the prospects for axiomatic truth theory. Nevertheless, axiomatic truth theory is adequate (*trafny*), as we learn in thesis C at [Tarski, 1983a, 266], [Tarski, 1986a, 184], [Tarski, 1933, 114].

Another addition to axiomatic truth theory Tarski considers is the ω-rule or "rule of infinite induction" [Tarski, 1983a, 259] that allows one to infer a numerical generalization from the (denumerably infinite) set of its instances. The rule was of significance due to its role in [Gödel, 1967b] and, as we will see, it played a major role in Carnap and Tarski's assimilation of Gödel's results; Tarski himself was considering its significance from at least 1926 onward [Tarski, 1983a, 260, note 1]. Tarski contemplates adding the rule to axiomatic truth theory as so far discussed, but notes that due to its "non-finitist nature ... it may well be doubted whether there is any place for the use of such a rule within the limits of the existing conception of the deductive method" [Tarski, 1983a, 260]. He then notes that the rule is intuitively valid in the sense that "it always leads from true sentences to true sentences" (a fact that will play a role later; § 7.4.2), and that adding it to axiomatic truth-theory remedies many of its weaknesses, e.g. the lack of categoricity [Tarski, 1983a, 261]. But, he cautions, it is not known whether axiomatic truth theory supplemented by the ω-rule is consistent; what is known, from Th. I, is that truth remains indefinable in syntax in the contemplated "deductive" (in an extended, non-finitary sense) theory. (This question gets settled, in a sense, in the Postscript, but see our remarks on the difference between the General Theory of Classes and Carnap's theory of levels in § 6.1.)

5.2 Inconsistency in everyday language

Tarski's views on the inconsistency of "colloquial language" have been the subject of incomprehension and derision for decades. We are now in a position to understand them. In § 4.1.3 we discussed the argument of § 1, focusing on the way in which straightforward employment of the "semantical definition" of truth would not have suited Tarski's purposes, since the semantical definition isn't formally correct. We can now look into the treatment of "colloquial language" in the section. Having gone through the difficulties involved in working out the semantical definition, Tarski segues into a remarkable conclusion:

> A characteristic feature of colloquial language (in contrast to various scientific languages) is its universality. It would not be in harmony with the spirit of this language if in some other language a word occurred which could not be translated into it; it could be claimed that 'if we can speak meaningfully about anything at all, we can also speak about it in colloquial language'. If we are to maintain this universality of everyday language in connexion with semantical investigations, we must, to be consistent, admit into the language, in addition to its sentences and other expressions, also the names of these sentences and expressions, and sentences containing these names, as well as such semantic expressions as 'true sentence', 'name', 'denote', etc. But it is presumably just this universality of everyday language which is the primary source of all semantical antinomies, like the antinomies of the liar or of heterological words. These antinomies seem to provide a proof that every language which is universal in the above sense, and for which the normal laws of logic hold, must be inconsistent. This

applies especially to the formulation of the antinomy of the liar which I have given on pages 157 and 158, and which contains no quotation-function with variable argument. If we analyse this antinomy in the above formulation we reach the conviction that no consistent language can exist for which the usual laws of logic hold and which at the same time satisfies the following conditions: (I) for any sentence which occurs in the language a definite name of this sentence also belongs to the language; (II) every expression formed from (2) by replacing the symbol '*p*' by any sentence of the language and the symbol '*x*' by a name of this sentence is to be regarded as a true sentence of this language; (III) in the language in question an empirically established premiss having the same meaning as (α) can be formulated and accepted as a true sentence.

If these observations are correct, then *the very possibility of a consistent use of the expression 'true sentence' which is in harmony with the laws of logic and the spirit of everyday language seems to be very questionable, and consequently the same doubt attaches to the possibility of constructing a correct definition of this expression* [Tarski, 1983a, 164–5].

The passage provokes a number of familiar complaints:

1. Natural languages can't be inconsistent because only theories can be inconsistent, and languages aren't theories [Putnam, 1975, 73], [Burge, 1984, 83–4], [Soames, 1999, 63].
2. Natural languages obviously aren't inconsistent since if they were communication in them would be impossible, but the very fact that Tarski's view can be stated shows that communication in natural language is possible [Sloman, 1971, 133–4].
3. Natural languages can't be inconsistent because an inconsistent language would be one in which contradictions were true, but if contradictions are true everything is the case, and, clearly, it is not the case that everything is the case [Soames, 1999, 55, 151].

In fact, as I will argue, Tarski's views on natural language are quite workable and perfectly sensible in view of his conception of meaning.

Tarski seems to take it as obvious what "everyday" or "colloquial" language is; presumably he has familiar "natural" languages like Polish and English in mind, but in fact his remarks fail to settle even this. As is standard, I will take it as understood that we can interpret his remarks in terms of such languages.

5.2.1   Inconsistent Kotarbińskian conventions

Despite the extent to which they have been pilloried, Tarski's views on everyday language are perfectly sensible in light of the conception of meaning and communication with which he worked around 1930. Recall the following points about Kotarbiński's conception of meaning:

- The utterance of a particular sentence directly expresses whatever the speaker intends it to express, along what we would now think of as broadly Gricean lines.
- A sentence indirectly expresses whatever the conventions of everyday language imply it would be used correctly directly to express.
- The concept intended to be expressed by a term is its connotation, the set of properties that would be grasped by someone who used it correctly in accord with the conventions directly to express a thought.
- There can therefore be a mismatch between the formal role determined by the conventions for a term and the concept it is intended to express.

This combination of views easily gives rise to Tarski's claims about the inconsistency of colloquial language. First, it is entirely possible for the conventions governing a language to be either inconsistent in themselves or inconsistent given contingent fact. This will happen precisely when the conventions involve expressions that are themselves bound up with thoughts about what the conventions require—that is, precisely when syntactic and semantic expressions are introduced.

We can see how this plays out entirely within Kotarbiński's conception by looking at a simple construction of the Grelling along his lines. Consider the following set of conventions:

1. Someone correctly accepts 'Xy' iff she thinks that 'X' applies to y and correctly accepts '$\neg Xy$' iff she thinks that 'X' does not apply to y. (Correctly accepting an atomic sentence requires thinking that the predicate applies to the objects denoted, while correctly accepting a negated atomic sentence requires thinking that the predicate does not apply to the objects denoted.)
2. Someone correctly accepts 'Axy' iff she thinks that x applies to y. ('A' is an application predicate.)
3. Someone correctly accepts 'Hx' iff she correctly accepts '$\neg Axx$' ('H' is a heterologicality predicate.)
4. Someone correctly accepts 'Xh' iff she thinks that 'X' applies to 'H'. ('h' names the heterologicality predicate.)

5. Either '$\neg p$' or 'p' is correctly accepted. (Bivalence: either a claim or its negation is always correctly accepted.)

These conventions determine that someone correctly accepts 'Hh' if and only if she thinks that 'H' applies to h, by (1), and hence, by (4) if and only if she thinks that 'H' applies to 'H'. By (2) and (3) they also determine that someone correctly accepts 'Hh' if and only if she thinks that h does not apply to h, and hence, by (4), if and only if she thinks that 'H' does not apply to 'H'. So someone will correctly accept 'Hh' if and only if she thinks both that 'H' applies to itself and that 'H' does not apply to itself.

So far we seem only have the unproblematic conclusion that nobody can correctly accept 'Hh' without having beliefs that imply a contradiction; all that would appear to follow from that is that nobody can accept 'Hh' without having beliefs that are true only if something both is and is not the case. However, parallel reasoning establishes that nobody can accept '$\neg Hh$' without having beliefs that imply a contradiction, either. By (1), someone correctly accepts '$\neg Hh$' if and only if she believes that 'H' does not apply to h, and hence, by (4), if and only if she believes that 'H' does not apply to 'H'. By (2) and (3), someone correctly accepts '$\neg Hh$' if and only if she thinks that it is not the case that h does not apply to h and therefore, eliminating the double negation, if and only if she thinks that h applies to h. By (4), such a person also believes that 'H' applies to 'H'.

Thus, '$\neg Hh$' is correctly accepted only by someone who has beliefs that can only be true if something both is and is not the case. However, by (5), one of either 'Hh' or '$\neg Hh$' is correctly accepted. If either is, though, a convention for conjunction introduction will get us an explicit contradiction being correctly accepted and a few more conventions for classical logic will get us that everything is correctly accepted. Since Kotarbiński and Tarski assume that the conventions governing the object language involve classical connectives, the above conventions specify a language in which every sentence is correctly accepted. Thus, if one believes what these conventions say one has beliefs that are true only if something both is and is not the case. Furthermore, the conventions are such that if one accepts them then one ought to conclude that all sentences are always correctly accepted.

Clearly, one might take this account of conventions to be overly simple; obvious emendations would make the conventions themselves inconsistent, and other obvious emendations, e.g. adding a notion of rejection,[7] would change the landscape in other pertinent ways. Our point here is that Kotarbiński and Tarski are working with a simple notion of correct acceptance that matches the conception of theoremhood in their understanding of a deductive theory (§ 1.4.3). The language governed by these conventions is "inconsistent" in the sense that anyone who agrees to the conventions is committed to logical falsehoods and, if we add classical logic, to every sentence's being correctly accepted.

We have now reconstructed Tarski's discussion of everyday language using only Kotarbiński's notions of a convention and indirect statement, and the account of correct assertion these conceptions entail. A language is inconsistent just in case its conventions jointly determine that both some sentence and its negation are correctly accepted; conventions that make some terms semantic and other terms syntactic, in the sense that they refer to expressions, are sufficient for this.

Let us now note a number of things that do *not* follow from the existence of inconsistent conventions on Kotarbiński's view. First, it doesn't follow that the intuitive meanings of expressions governed by these conventions, that is, the concepts expressed by them, are themselves somehow "inconsistent". The concept expressed by a term is its connotation, a set of properties. 'A' connotes the property that two things have when the first applies to the second. The thought that one has if and only if one correctly accepts 'Axy' is that the ordered pair (x,y) has the application property. There may well be such a property even if the conventions governing some language in which 'A' appears are jointly inconsistent. The existence of inconsistent conventions governing 'A' and related expressions are perfectly compatible with the claim that speakers grasp the concept that 'A' is intended to connote, and even with the claim that this concept is somehow simple and obvious.

Second, it doesn't follow that any individual convention on its own is subject to criticism. The application property may well be such that an expression that connotes it and which has a role in a language that captures this connotation must be governed by convention (2) above; likewise, the property of heterologicality may be such that any expression that connotes it must be governed by (3). None of these conventions need be flawed on its own, and, indeed, each may be required for adequate expression of intuitive meaning. Yet it may still be the case that the conventions cannot coherently be combined—which is to say that there may be no such thing as a logically non-trivial language in which every concept can be expressed. Taken correctly, this is simply a corollary of the standard limitative results of the period [Patterson, 2009, 414].

Third, it doesn't follow that those who use these conventions do not express determinate thoughts in particular assertions. Since speakers' acceptance of the conventions isn't closed under logical consequence, they might accept a set of conventions that in fact imply that every sentence is assertible in every circumstance without also accepting that every sentence is assertible in every circumstance. What speakers accept is some finite set of claims about which sentences express which thought-constituents, and these speakers will in general apply their beliefs about this only to the point of their most immediate implication of some claim about what thought a given sentence is correctly used to express. Speakers as a group, then, will converge on beliefs to the effect that various sentences are correctly asserted only by those who have certain beliefs, and such speakers will, in using those sentences, directly express (to use Kotarbiński's terminology) precisely those beliefs in the use of those sentences—even though, unbeknownst to them, the conventions they accept don't actually determine that particular thoughts are indirectly expressed by particular sentences.

Fourth, speakers do not need to be especially dim-witted to end up speaking an inconsistent language. Speakers who agree to conventions that are in fact inconsistent in this sense without noticing it can go about their business pretty much unhindered. Assertion of a particular sentence taken to be correct by fellow logically ignorant adherents to the conventions will lead them to attribute a determinate thought, and if one is in fact using the sentence as intended (that is, directly expressing with it that thought which the conventions say that it indirectly expresses), it will be exactly the thought one directly expressed in the assertion. So communication, as Kotarbiński construes it, can happen unimpeded by the fact that the conventions of a language jointly reduce it to logical triviality.[8]

Finally, it isn't any part of Kotarbiński's conception to maintain that contradictions are true. The view is only that contradictions are correctly asserted according to the conventions of language that speakers actually accept. There is no more mystery in this than there is in the idea that any other system of rules or conventions that people accept might be such that its statement is such that one could actually abide by the rules only if everything were the case.

Now such speakers, if they get to trying systematically to work out lines of thought by tracing out what has to be correctly assertible given that other things are, given the conventions, may get themselves into trouble, and they may notice that there is a problem. But the business of expressing thoughts in particular situations may proceed unimpeded at no cost. Furthermore, accepting inconsistent conventions doesn't require anyone to believe contradictions; the conventions of course determine that certain setences both are and are not correctly asserted, but it isn't a condition of accepting the conventions themselves that one must recognize this consequence.

5.2.2   Tarski after Kotarbiński

None of the usual complaints about Tarski's notion of an inconsistent language applies to the Kotarbińskian view. We can now make short work of understanding Tarski's notorious remarks. An inconsistent language is one the conventions governing which imply that all of its sentences are correctly asserted.[9] As we saw earlier, the basic notion is not truth but correct assertability; hence, Tarski does not mean by an inconsistent language a language in which contradictions or all sentences are true— contrary to [Soames, 1999, 55] and in accord with [Ray, 2003, 71] and [Patterson, 2006, 157]. When Tarski writes that:

> If these observations are correct, then *the very possibility of a consistent use of the expression 'true sentence' which is in harmony with the laws of logic and the spirit of everyday language seems to be very questionable, and consequently the same doubt attaches to the possibility of constructing a correct definition of this expression* [Tarski, 1983a, 164–5].

he means that given the content "true sentence" is intended to express, the expression of which requires the T-sentences to be assertible by the conventions governing the language, introducing such conventions into a language in which (a) the connectives are classical, and (b) "the spirit of everyday language" is maintained in the sense that it is maintained that every intuitive concept (and so, for instance, syntactic concepts denoting expressions of the language can also be expressed) will render that language inconsistent in the sense adumbrated above: the conventions jointly will determine that every sentence is assertible.

We can also understand why it doesn't follow, on Tarski's view, that communication is impossible in everyday language: given Kotarbiński's accounts of convention, indirect expression and communication, communication is possible in an inconsistent language. It likewise doesn't follow, for reasons stated above, that only irrational people can speak inconsistent languages. Finally, then, it doesn't follow that the very fact that Tarski spoke, wrote and was understood in everyday language is a refutation of his view. Quite the contrary: the overall view explains pefectly well how its own truth is compatible with Tarski's being able to do so.

As noted in places above, Tarski's view is not that the concept of truth itself is "inconsistent" - or, more accurately, that the content of the concept of truth is "inconsistent" - ([García-Carpintero, 1996, 125–6] [Soames, 1999, 51] and many others). The concept has a perfectly clear content, stated in the "semantical definition". Languages, governed by conventions, can be inconsistent in the sense that the conventions require the assertion of every sentence, and the paradoxes show that "colloquial" languages are like this. Note that the view that Tarski holds that there can be inconsistent concepts is very naturally allied to the common misconception that each defined Tarskian truth-predicate expresses a different concept (§4.3.2): if "true" in English expresses an "inconsistent" concept, but a defined "true" for some formalized language expresses a consistent concept, then different truth predicates express different concepts. We see here how various aspects of the standard reading of Tarski reinforce one another.

We have here a reading on which Tarski's notion of an inconsistent language does not involve the claim that an inconsistent language is one in which contradictions are true. An inconsistent language is simply one governed by conventions that imply that contradictions may correctly be asserted. As we noted with Kotarbiński, nothing untoward follows from this: attempts to lay down rules in any domain can misfire, and Tarski's claim is simply that the rules of use that speakers of natural language attempt to implement have this flaw. Soames [Soames, 1999, 64] responds to this with the objections (1) that "it is hard to imagine that it should be a condition of my speaking English that I be willing to assert things that are not true", and (2) that "Tarski nowhere explicitly states that a language can dictate the assertability of a sentence that is not true". In response to (2) we can note, first, that Tarski nowhere says the opposite, either, and that the conception of deductive systems in articles like [Tarski, 1983g] and [Tarski, 1983c] obviously allows for the possibility of inconsistent rules. As for (1), we need to be careful with the notion of "speaking English". If to speak English is to do everything the rules require, then Tarski's view implies that it is impossible to speak English—but on this definition of "speak English" the view is quite reasonable, and it doesn't imply that it is a condition of speaking English except in the trivial sense that (given classical logic) every condition is a necessary condition of doing the impossible. If we take the more reasonable view that speaking English is a matter of intending to abide by its conventions, then people do speak English, but by the above considerations on how rational people can speak inconsistent languages, it doesn't follow that it is a necessary condition of intending to abide by the conventions of English that one be willing to accept explicit contradictions or other obvious falsehoods; it follows only that one can only do what one intends to do if one does so.

In § 5 of [Patterson, 2006] I extended similar remarks to the view that Tarski holds that no languages are semantically closed, though some appear to their speakers to be. Like the rest of that paper, though the point still seems to me to be essentially right, it is vitiated by my adherence at the time of writing to more standard views of Tarski's philosophy of language on which Tarski was really only concerned with introducing an expression that had as its extension the set of truths of a language. It is correct, as I claimed there, that no language is universal in the sense of having expressions that define absolutely all sets, or even universal in the sense of having expressions that define all of the subsets of the set of their own sentences.[10] We can however also say, as was not said there, that natural languages *are* "universal" in the sense of intuitive meaning: for every intuitive concept, natural language can be extended to include an expression governed by a convention sufficient for sentences involving it indirectly to express thoughts involving that concept (contrary to the typical response to the modal objection discussed above, as we noted, Tarski allows that languages can be extended [Tarski, 1983a, 164]). In principle, everything can be indirectly expressed according to the conventions of natural language, and, more obviously, everything can be directly expressed with a sentence of natural language, given the Gricean story in the background of Tarski's Kotarbińskian views. In that sense, natural languages are universal; the problem is that they buy this universality at the price of formal incoherence and total semantic indeterminacy: strictly speaking, taking the rules seriously, every sentence of a natural language is assertible: just construct a semantic paradox and reason by *ex falso* from there. Moreover, precisely because everything is assertible, no predicate can be assigned any particular extension as opposed to any other.[11]

I stand by another contention of that article, as should be clear from the above: Tarski doesn't in any serious sense propose accounting for the semantics of natural language by imposing a "Tarski hierarchy" on it. As he writes, this would involve "the thankless task of a reform of this language" [Tarski, 1983a, 267]. Since doing so is required to pursue the semantics of colloquial language by "exact" methods, we can conclude that Tarski doesn't hold the informal account of such language in terms of convention in Kotarbiński's style to be exact, but in the body of CTFL he clearly follows Kotarbiński in holding that the correct account of meaning in natural language is in terms of convention and intention, not extensional semantics.

6
Transitions: 1933–1935

Well past 1931 semantics was for Tarski simply a tool for producing explicit definitions of certain concepts that appeared both suspect and difficult to avoid: truth, denotation, satisfaction and (semantic) definition. Some of the tools thereby developed were even mathematically fruitful, as shown in the later pages of ODS and its sequel, "Logical Operations and Projective Sets", co-authored with Kuratowski. But it seems not to have occurred to Tarski to think of meaning itself in terms of semantic word–world relations; he was too wedded to the conception of meaning in terms of the expression of thought and the constraints placed on this by the deductive structure of an axiom system. His willingness to present "Some Methodological Investigations" in 1934 and publish it in 1935 shows how far he was, even that late, from thinking that semantics could stand on its own or take the place of the older meaning-theoretic notions he had acquired from Kotarbiński and Leśniewski. Padoa's method clearly should be construed in semantic terms, yet in 1934 Tarski is willing to re-use material from 1926 in which the notion of an interpretation of a term receives the same psychologistic treatment it gets from Padoa himself.

Given the timing and the nature of the change, especially as it will be discussed in the next chapter with respect to Tarski's conception of logical consequence, one impetus for the shift in Tarski's thinking has to be his careful reading of Carnap's *Logical Syntax of Language* and his interaction with Carnap himself. In brief: seeing what goes wrong in Carnap's definition of "analytic" for Language II led Tarski to see that what was missing was his semantics. This, in turn, in the space of about a year led Tarski to shed the older conception of meaning with which his project began and to move to a more contemporary conception of truth-theoretic semantics. I begin with one text that shows the influence of *Logical Syntax* to some extent, but also shows Tarski moving away from his Polish roots, especially with respect to Leśniewski, in repudiating the allegiance of the main text of "The Concept of Truth" to the theory of semantical categories and hence to STT as the true system of logic: the 1935 Postscript to the German translation.

6.1   The 1935 postscript

In the body of CTFL Tarski subscribed to an interpretation of STT in terms of Leśniewski's theory of "semantical categories". For our purposes here it suffices to note that the system assumes that any given functional category determines the categories of its arguments—so, in particular, there is no such thing as a functional category that can take either names or predicates as arguments and form a sentence. At that time, Tarski wrote:

> the theory of semantical categories penetrates so deeply into our fundamental intuitions regarding the meaningfulness of expressions, that it is scarcely possible to imagine a scientific language in which the sentences have a clear intuitive meaning but the structure of which cannot be brought into harmony with the above theory [Tarski, 1983a, 215].

This commitment shapes his discussion of the possibility of a definition of truth for a language of infinite order, implying as it does that for such languages truth cannot be explicitly defined, cannot therefore reductively be eliminated in favor of unproblematic expressions from the "morphology of language", and, concomitantly, that no expression of the metalanguage can semantically define the set of truths of the object language, dooming consistent axiomatic truth-theory to a failure of provable categoricity. The upshot of the whole discussion is that as applied to the most interesting and powerful languages, the intuitive concept of truth can be only imperfectly expressed.

Tarski relaxes his adherence to the theory of semantical categories in the postscript:

> Today I can no longer defend decisively the view I then took of this question. In connexion with this it now seems to me interesting and important to inquire what the consequences would be for the basic problems of the present work if we included in the field under

consideration formalized languages for which the fundamental principles of the theory of semantical categories no longer hold [Tarski, 1983a, 268].

Note the hesitation here: Tarski cannot defend the view "decisively"— hardly a wholesale rejection. Relatedly, notice that Tarski doesn't quite repudiate the view that no language of which the fundamental principles of the theory of semantical categories fail to hold can have sentences with clear intuitive meanings. Rather, the basic problems of the work will be considered with respect to languages of which these principles do not hold. It is often claimed (e.g. [Frost-Arnold, 2004, 272], [Feferman, 2008, 86], [Creath, 1999, 69]) that Tarski *changes* his view in the Postscript and comes to hold that truth can be defined for the General Theory of Classes using expressions of infinite order, as though the shift in the Postscript were merely a matter of adding some transfinite levels atop the hierarchy of STT. This isn't accurate: Tarski doesn't change his views about GTC formulated in STT. Rather, he widens the scope of his considerations to languages that don't adhere to the theory of semantical categories and then points out that for *those* languages a truth definition is available because expressions that are, by *their* standards, "of infinite order" are available. It remains the case that truth can't be defined for GTC in a language that adheres to the strictures of the theory of semantical categories.[1] (Admittedly, Tarski himself isn't helpful on this point. Theses A and B of § 6 [Tarski, 1983a, 265] aren't actually in contradiction with Theses A and B of the Postscript, since the shift from the theory of semantical categories to Carnap's theory of levels induces an ambiguity in "order".)

Now, one might think that since the basic problems of the work had everything to do with intuitive meaning, the focus is still squarely on that. The Postscript, however, is silent on the topic of intuitive meaning, making simply the formal point that if one allows expressions of infinite order, then for every language there is a language of higher order and hence for every language there is a metalanguage in which an explicit truth-definition can be given. What is in the offing is not so much an extension of previous questions about intuitive meaning as a waning of interest in them. If Tarski had remained interested in the Leśniewskian views about meaning that had animated him in the first place, he would have remarked on how the abandonment of the theory of semantical categories affected the conception of meaning.

The shift that matters to the results highlighted in the Postscript is Tarski's allowing that the notion of an expression of transfinite order can be countenanced. The larger shift is allowing expressions that don't fix the orders of their arguments:

> Since the principles of the theory of semantical categories no longer hold, it may happen that one and the same sign plays the part of a functor in two or more sentential functions in which arguments occupying respectively the same places nevertheless belong to different orders [Tarski, 1983a, 269].

Since Tarski wants to adhere to the principle that the order of an expression is the least ordinal greater than any that specifies the order of any argument it takes, but there is no finite ordinal α such that ω is the least ordinal greater than α, the only way to get expressions of transfinite order is to have expressions that take arguments of *all* finite orders and hence to allow variability in the order of the arguments that a functional expression takes (cf. [Sundholm, 2003, 118]). It is in this shift that Tarski abandons the theory of semantical categories and hence simple type theory as he had construed it. As we have noted in § 4.2.2 the constraints of STT had long been an irritant that Tarski salved with "systematically ambiguous" formulations anyway.

What is motivating the shift away from the theory of semantical categories of which Tarski had been so certain only a few years earlier? The answer comes at footnote 2 to [Tarski, 1983a, 270]: Tarski has the theory of levels from Carnap's *Logical Syntax* in mind. So the Postscript is written, really, to relax the theory of semantical categories down to the more liberal constraints of the theory of levels of § 53 of *Logical Syntax*, and this theory allows both expressions of infinite order and predicates and functors that take arguments of variable order [Carnap, 2002, 188–9]—e.g. the identity predicate gets assigned order ω because it takes arguments of every finite order.[2]

The abandonment of the theory of semantical categories then allows that truth can explicitly be defined for a language of any Carnapian order, finite or transfinite, and Tarski correspondingly modifies the conclusions he drew in the original text. The original theses A and B stated that truth can be defined for a formalized language of finite order, but not one of infinite order [Tarski, 1983a, 265], while the revised theses state that truth can be defined for every formalized language in a metalanguage of higher order, be it finite or transfinite [Tarski, 1983a, 273]. The original thesis C stated that for a language of infinite order "the consistent and adequate" (*konsequenter und richtiger, konsekwentnie i trafnie*) use of the concept of truth was possible on the basis of an axiomatic truth theory. As Tarski notes, it loses its importance since axiomatic truth-theory done in a language not of higher order than the object language becomes a mere curiosity when truth for the same object language can be defined in a metalanguage of higher order. Concomitantly, the open questions about the consistency of axiomatic truth-theory augmented with the ω-rule, are settled: since truth can consistently be defined, and since the axiomatic theory with the ω-rule is a sub-theory of the theory including the definition, the extended axiomatic theory is consistent [Tarski, 1983a, 273].[3]

We should note that in addition to relaxing the theory of semantical categories down to Carnap's theory of levels, the Postscript considers a shift to first-order axiomatic set theory as well [Tarski, 1983a, 271 note 1]. The main points in the postscript carry over: one can define truth for a first-order set theory only with a more powerful set theory. In this form the Postscript's treatment of the issues more closely resembles a contemporary take on "Tarski's Theorem": arithmetic truth is not arithmetically definable, though, as noted, the explicit focus in the Postscript is still on explicit definition and not semantic definition.[4]

What do we see in the Postscript that is of most interest to our study here? The most important shift is the weakening of the hold on Tarski of the views about meaning he inherited from Leśniewski, and a diminution of interest in the question that got him started in the first place of how intuitive semantic concepts are to be formally expressed. The Postscript, for instance, is almost silent on what happens to intuitive meaning when the theory of semantical categories is scrapped in favor of Carnap's theory of levels. All Tarski gives us is the remark that he no longer "defends decisively" the theory of semantical categories. Since at this date it is unlikely that Tarski has ceased to hold that the problem of defining truth requires sentences to have "concrete, intelligible meanings" [Tarski, 1983a, 167], all we can conclude is that Carnap has convinced Tarski that it isn't obvious that expressions within the theory of levels don't have such meanings, contrary to Tarski's emphatic endorsement of the theory of semantical categories at [Tarski, 1983a, 215].

Now given the tight connection in Leśniewski and therefore Tarski's thought between the capacity of language to express thoughts and that language's being structured in accord with the theory of semantical categories, it follows from the Postscript's apostasy from that theory that the Postscript has little to say about the effect on the intuitive thought expressed by language of the relaxation of the theory of semantical categories down to Carnap's theory of levels. With increasing tolerance comes a decreased emphasis on the idea that logic expresses thought (§1.4.2). When language is viewed as expressing thoughts that stand in a determinate set of deductive logical relationships, logic is the attempt to capture exactly these relationships and no others. In order, by contrast, to find room in one's view for multiple logical systems that are all equally "right", this picture needs to be replaced by one of various other familiar ones—e.g. that of the rules of various games that can be played, on a more conventionalist view, or a thoroughly empiricist view of the sort to which Tarski was ultimately attracted [Frost-Arnold, 2004], [Mancosu, 2005]. But in the Postscript we see that the basic expressive conception of language from within which his project began was losing its grip on Tarski, and that he was moving to a more open-minded, Carnapian, "tolerant" conception of alternative logical systems.

In other respects, the Postscript remains squarely within the earlier period. Explicit definition that implies the T-sentences remains the central concern, with the definability of sets only in the background and largely relegated to a sketchy footnote on first-order axiomatic set theory. The conception of logical consequence is still fully derivational: consequence is to be defined as the closure of some axioms under some transformation rules [Tarski, 1983a, 269]. Though the older conception was losing its grip a bit at the time of the writing of the Postscript, the most significant changes were yet to come.

6.2   Carnap on analyticity and truth

As is often noted, Tarski studied [Carnap, 2002] carefully and is thanked for suggestions and corrections in many places throughout the 1937 expanded English edition. When he studied *Logical Syntax* Tarski would have found two passages that were close to his interests with respect to the topics that concern us here: the argument of § 60b that truth cannot be defined in syntax, and the definitions of 'analytic' and 'contradictory' in § 34.[5] By Tarski's standards the former is in error and the latter involves convolutions that Tarski's conception of truth would strip away.

The argument of [Carnap, 2002, § 60b] purports to show, first, that truth cannot be defined for a language sufficient to express its own syntax in that language itself and, second, that even in a language of greater expressive power, truth cannot be defined. The argument for the first claim runs via a familiar demonstration that introducing a predicate such that the T-sentences are theorems into a language S that can formulate its own syntax will generate sentences that attribute untruth to themselves [Carnap, 2002, 214–6]. Carnap's official formulation involves a version of the "visiting card paradox" involving two sentences, the first of which says that the second is false while the second says that the first is true in order, as Carnap explains, to avoid the impression that self-reference is somehow the problem. So far this is perfectly standard.

The argument for the second claim is where things come off the rails. Carnap writes:

> This contradiction only arises when the predicates 'true' and 'false' referring to sentences in a language S are used in S itself. On the other hand, it is possible to proceed without incurring any contradiction by employing predicates 'true (in S_1)' and 'false (in S_1)' in a syntax of S_1 which is not formulated in S_1 itself but in another language S_2. S_2 can, for instance, be obtained from S_1 by the addition of these two predicates as new primitive symbols and the erection of suitable primitive sentences relating to them, in the following way: 1. Every sentence of S_1 is either true or false. 2. No sentence of S_1 is at the same time both true and false. 3. If, in S_1, $\mathfrak{S}_2$ is a consequence of $\mathfrak{K}_1$, and if all sentences of $\mathfrak{K}_1$ are true, then $\mathfrak{S}_2$ is likewise true. A theory of this kind formulated in the manner of a syntax would nevertheless not be a genuine syntax. *For truth and falsehood are not proper syntactical properties*; whether a sentence is true or false cannot generally be seen by its design, that is to say, by the kinds and serial order of its symbols [Carnap, 2002, 216].

Carnap goes on to propose that this doesn't matter because in the syntax language for S_1 all the work one would have wanted to do with 'Sentence A is true' can be done either by A itself or by 'A is analytic'.

The problem here is Carnap's italicized sentence and the explanation that follows it. For 'true' to be definable in syntax and hence, ultimately, in arithmetic it needn't be the case that it is 'syntactic' in the sense that it is a property of a sentence determined by the kinds and serial order of its symbols; it need only be the case that it can be defined in terms of expressions that apply to sentences in virtue of the kinds and order of their symbols. Coffa, at [Coffa, 1987, 566–7] notes the problem, but gives only a partial diagnosis:

> In general, whether a sentence is true or false cannot "be gathered" through the techniques available in syntax. Notice that whether a sentence is analytic or not (in Carnap's sense) cannot be gathered from the techniques available in Carnap's syntax; and, more to the point, whether a sentence is true or false cannot be gathered by the techniques available in semantics either. What could have led Carnap to think that whether a concept C is definable on the basis of certain techniques depends on whether those techniques allow us to identify

the instances of C? The obvious answer is: a verificationist prejudice [Coffa, 1987, 567].

It may well be right that that verificationism encourages the thought that if truth can be defined in terms of syntax then syntax must decide whether any given sentence is true, the idea being that the meaning with which any predicate is endowed must be associated with a verification procedure. The problem Coffa alludes to is that "analytic" itself doesn't apply to a sentence merely in virtue of the "kinds and serial order of its symbols", since analyticity depends on whether various substitution-variants of a sentence are consequences of everything. Carnap later described his mental block at the time this way:

> When Tarski told me for the first time that he had constructed a definition of truth, I assumed that he had in mind a syntactical definition of logical truth or provability. I was surprised when he said that he meant truth in the customary sense, including contingent factual truth. Since I was only thinking in terms of a syntactical metalanguage, I wondered how it was possible to state the truth-condition for a simple sentence like "this table is black". Tarski replied: "This is simple: the sentence 'this table is black' is true if and only if this table is black" [Carnap, 1963, 60].

Once the object language was included in or translated into the metalanguage, one no longer had to cast about within the syntactic and proof-theoretic resources of the metalanguage to find the terms in which to construct the definition of truth. As long as the object language itself was taken to be scientifically legitimate, there was, as Tarski saw, no reason not to make use of its expressive resources in the metalanguage as well.

Turning to our second topic, Tarski, already armed with his procedures for explicitly defining truth, would immediately have appreciated how unnecessarily complicated Carnap's definition of analyticity for Language II was, while it nevertheless was, in its essential structure, quite close to Tarski's own way of doing things. This point about the relation between Carnap and Tarski is familiar from the literature; see e.g. [Coffa, 1987] and [de Rouilhan, 2009].

Analyticity is defined for Language II in terms of *valuations*. Carnap first points out that although we can, in fine syntactic form, take the valuation of a first-order variable to be a denumerably infinite set of names (we see here the same understanding of the values of a variable as we found (§3.1.1) in the 1929 *Abriss*) and so understand a sentence open on a first-order variable as analytic when every sentence that results from substituting a name for the variable is, when it comes to higher-order quantification this method won't give the result we want, for the (now) obvious reason that "as a result of Gödel's researches it is certain, for instance, that for every arithmetical system there are *numerical properties which are not definable*" [Carnap, 2002, 106].

With only denumerably many symbols to work with Carnap makes the striking and desperate move of associating higher-order variables with set-theoretic constructs out of names themselves, letting second-order (one-place) predicate variables range over the power set of the set of names—thus buying himself more than denumerably many items for a second-order variable to range over [Carnap, 2002, 107]. Despite the ascent to the transfinite somehow Carnap convinced himself that this was still properly "syntactic". With sufficient numbers of interpretants for higher-order variables secured in this manner, Carnap can then proceed in what is (following Tarski) the obvious way—a universal generalization is analytic just in case all of its instances are, etc. [Carnap, 2002, 107ff]. The issue, however, is how "analyticity" is defined for the result of substituting n appropriate constructs out of the set of names to n open positions in some open sentence. Here is what Carnap writes with respect to the closure of a second-order one-place open sentence:

> Now if $\mathfrak{B}_1$ is a particular valuation for 'F' of this kind, and if at any place in $\mathfrak{S}_1$ 'F' occurs with $\mathfrak{St}_1$ as its argument (for example, in the partial sentence '$F(0'')$'), then this partial sentence is—so to speak—true on account of $\mathfrak{B}_1$, if $\mathfrak{St}_1$ is an element of $\mathfrak{B}_1$, and otherwise false. Now, by the evaluation of $\mathfrak{S}_1$ on the basis of $\mathfrak{B}_1$, we understand a transformation of $\mathfrak{S}_1$ in which the partial sentence mentioned is replaced by $\mathfrak{N}$ if $\mathfrak{St}_1$ is an element of $\mathfrak{B}_1$, and otherwise by $\sim \mathfrak{N}$. The definition of 'analytic' will be so framed that $\mathfrak{S}_1$ will be called analytic if and only if every sentence is analytic which results from $\mathfrak{S}_1$ by means of evaluation on the basis of any valuation for 'F'. And $\mathfrak{S}_1$ will be called contradictory when at least one of the resulting sentences is a contradictory sentence. We shall lay down analogous rules for other $\mathfrak{p}$-types [Carnap, 2002, 107].

As Tarski was well in a position to appreciate, the whole procedure makes no sense at all at the crucial point unless one keeps the intuitive notion of truth which is supposedly being eliminated in mind. This isn't to accuse Carnap's definition of circularity—this would be analogous to the mistaken criticism of Tarski we rejected above in § 4.3.1. The point, rather, is that simply admitting that the notion of truth is in play makes all of the complications unnecessary. Why should the fact that a is in a set of names associated with F allow us to "transform" Fa into '$0 = 0$' rather than '$0 \neq 0$'? After all, in themselves, we just have a bunch of names here. The only reason for the procedure is that we tacitly allow—as Carnap can't restrain himself from mentioning—that a name's being in a set of names stands in for the idea that what is named is in the set of things that satisfy the predicate and that the sentence is thereby *true* (cf. [Creath, 1999, 69]).

So what Tarski found in §§ 60 and 34b was that, on the one hand, Carnap had no good argument against the possibility that semantic expressions be defined in syntax, and that, on the other, Carnap's own procedure with respect to Language II appealed to reference, satisfaction and truth in everything but name. The only thing preventing Carnap from seeing the fruitfulness of the truth-definitions he had nearly arrived at himself was a lingering prejudice against semantic notions backed up by the oversight of § 60 on which a properly scientific metalinguistic treatment of a language couldn't involve the terms of that language in addition to its own proper syntactic terms.

Before looking at the actual developments in the next chapter, we should pause now to understand the shift in the offing once one realizes that Carnap's definition of analyticity is best replaced by a conception based on Tarski's semantics. As long as consequence is conceived of in terms of primitively valid transformation rules, syntactic and proof-theoretic notions are basic to one's conception of logic. Even with semantics added in later in the style of [Tarski, 1983f], proof-theoretic consequence is needed before extensions can be assigned to expressions, since the set of provable theorems as a whole is what constrains their semantic interpretation.

By contrast, if we take the extensional semantic values assigned to primitive terms (be they objects, sets, n-tuples, or various functions therefrom and thereto) as basic and grant ourselves merely formation rules, a notion of consequence drops out of the account for free: a sentence is a consequence of others relative to a certain fixed vocabulary if and only if every semantic assignment to the remaining expressions either makes the former true or one of the latter false. Of course, in place of wrangling over which transformation rules are valid here one gets instead the now-familiar question of the logical constants: which selection of fixed vocabulary, if any, produces an account of *logical* consequence? The primary question to be addressed when it comes to the account of logical consequence developed in this way is whether it constitutes any advance over the account in terms of transformation rules. But the point now is that once one has the idea of defining analyticity and more generally consequence in terms of truth, one sees that one can *judge* the transformation rules previously treated as primitively valid by the standards of the resultant account of consequence.

Overall, then, Tarski's encounter with *Logical Syntax* allowed him to see that Carnap's treatment of analyticity really appeals to intuitive semantic notions anyway, that Carnap has no argument against the inclusion of semantic notions in serious theory, and that once Carnap's treatment of analyticity is replaced with Tarskian semantics, everything can be done without the transformation rules.[6]

Tarski began with a derivability conception of consequence associated with the Intuitionistic Formalist idea that a deductive theory was primarily a vehicle for the expression of thought and reasoning. But this could only be taken so far; the convolutions of Carnap's definition of analyticity for Language II shows why it ultimately can't be worked out. And then the semantics he developed along the way as a subsidiary tool turned out to be just what was required to fix the problem. The result was truth-theoretic semantics and the definition of consequence as we know it.

6.3   The establishment of scientific semantics

Having set out to establish that his semantics should replace syntax in Carnap's system, Tarski was led to see what semantics could have been doing that so far he wasn't doing with it. We have seen that in 1934 Tarski still understood Padoa's method in Intuitionistic Formalist terms and ultimately psychologistically, and that as late as the Postscript to CTFL Tarski showed no inclination to define logical consequence semantically rather than in terms of a list of intuitively valid transformation rules. But once Tarski set himself to thinking about what he could accomplish with his techniques for defining semantic terms, within a few months he had cast aside the old way of understanding consequence and with it, though less explicitly, the conception of meaning from which he had begun. We will look at the first point in the next chapter and the second in the chapter after that.

Tarski could only have wanted to let Carnap and anyone who followed him know how much more effectively semantics would serve the purposes to which Carnap was putting the convolutions of his "syntax". The fog in the definition of analyticity for Language II could be cleared by a straightforward definition of truth in acceptable terms of just the sort that Tarski was in a position to provide. Tarski got his chance in discussions with Carnap, who was quickly convinced, and who encouraged Tarski to present his ideas at the planned September 1935 Unity of Science congress in Paris [Carnap, 1963, 61], [Feferman and Feferman, 2004, 95].

Tarski arrived at the International Conference of Scientific Philosophy in Paris with two papers to present: "The Establishment of Scienfic Semantics" and "On the Concept of Logical Consequence". The first is more or less an advertisement for Tarski's method of defining truth, while the second represents an essential extension of semantics to a domain previously relegated to the realm of syntax and intuition, the consequence relation. We will cover the first briefly here. The second will be our topic in the next chapter.

In [Tarski, 1983b] Tarski reviews the content of CTFL and, for the audience of the Congress, tries to make things as exciting as possible. The first few paragraphs read like the introduction to "On the Concept of Truth": semantic concepts are intuitively clear but attempts to make rigorous sense of them bog down in paradox. The solution is rigorously to distinguish object and metalanguage and to accept the relativization of the definitions to come to one object language at a time. We set up a formal syntax in the metalanguage, include translations of the object language's distinguished vocabulary, and then define truth in the style of § 3 of CTFL. The article closes with a brief recapitulation of the remarks on Gödel from the Postscript.

Missing in all of this is much philosophizing about the concept of truth of the sort that had animated the project just four years earlier. A single sentence in the concluding paragraph gives a nod to the philosophical ambitions, but the overall pitch of the article is to the "scientific" philosophers in the audience. Interestingly, Tarski doesn't anticipate the definition of consequence set out in the other lecture. To this we turn in the next chapter.

7
Logical Consequence

As with his account of truth, Tarski's account of logical consequence has come in for a great deal of abuse. Characteristic here is Etchemendy's familiar assessment that Tarski's "account of logical truth and logical consequence does not capture, or even come close to capturing, any pretheoretic conception of the logical properties" [Etchemendy, 1990, 6]. Tarski's conception of consequence has likewise been subject to "defenses" that either attempt to make virtues out of its least attractive features or are implausible as readings of what Tarski actually said. Many of these defenses read "On the Concept of Following Logically" through the lens of the lecture "What are Logical Notions?", a lecture delivered thirty-one years after the original presentation of the conception of consequence. Though there is some basis for this in work contemporary with CLC, e.g. [Tarski, 1983i], the astonishing fact about the literature is that pretty much nobody has bothered to follow up the article's many references to Carnap's *Logical Syntax* in any detail—this despite the fact that the only conception of consequence Tarski mentions is Carnap's. I will argue that the article intends nothing more than an improvement on Carnap's conception that consists in replacing Carnap's reliance on the notion of substitution with Tarski's treatment of semantics, and transformation rules taken as primitive with the notion of truth-preservation leaving everything else untouched, and that all of the important claims and arguments in the article are directed at Carnap. In particular, Tarski's conception of formality is not the one extractable from the later article, on which logic is "formal" in the sense of being concerned with mathematical structure, but is rather nothing other than Carnap's notion of syntactic form. Reading Tarski in terms of Carnap will allow us better understanding of many passages of the article that have caused difficulty and it will also allow us to highlight just where the implausibility of the conception lies.

7.1 Tarski's definition

7.1.1 Synopsis

We can begin by briefly summarizing the text. Tarski announces the aim of capturing the content of the concept of logical consequence (or "following logically", as [Tarski, 2002] has it), with caveats about the "everyday, 'pre-existing'" usage of the term expressing it being unclear, the intuitions about it being "murky", and so forth—compare here the opening remarks in ODS and CTFL. Here the aim seems to be in accord with Intuitionistic Formalism, in that the aim is to capture the content of an ordinary concept—but that we are now at some remove from Intuitionistic Formalism is indicated among other things by the fact that Tarski makes no heavy weather over explicitly characterizing the object language and the metalanguage. In part this is due to the generality inherent in the case: Tarski will show, given a definition of satisfaction, how to define a model, and in terms of that, consequence. But it is also due to the fact that assignments of semantic value will be made to do the work previously done by transformation rules.

He then surveys the familiar syntactic conception of consequence to which he himself, as we have seen, subscribed, with reservations [Tarski, 1983a, 252], as late as the Postscript to CTFL, and notes that Gödel's results show that it doesn't capture the "everyday intuitions" [Tarski, 2002, 178] about what follows from what.[1] In particular, as Gödel had recently shown, syntactic conceptions of consequence are ω-incomplete [Tarski, 2002, 177–8]. The ω-rule, which allows the derivation of a universal generalization over numbers from all of its numerical instances fixes this but isn't finitary [Tarski, 2002, 178]. Using Gödelian techniques allows the formation of finitary rules that allow the derivation of universal generalizations from coded claims to the effect that all of their instances are provable [Tarski, 2002, 180], but by the same Gödelian results the result is still ω-incompleteness [Tarski, 2002, 181]. More generally, Gödel sentences for a theory follow logically, according to Tarski, from this theory "in the everyday sense", but for any proposed set of rules, including finitary coded replacements for the ω-rule, there is some such sentence that isn't provable.

Tarski next notes Carnap's attempt at capturing the "proper concept" of logical consequence, faulting Carnap's definition for Language II as "too special and complicated" [Tarski, 2002, 182], his definition for Language I as being unsuited for extension to "less elementary" languages [Tarski, 2002, 192], and his definition in General Syntax as making the "denotation" of the defined concept "dependent in an essential way on the richness of the language which is the object of consideration" (this is the issue with Condition (F) and Tarski's modification of the account; see below), and as depending on a basic concept of consequence in terms of stipulated transformation rules that Carnap leaves as primitive [Tarski, 2002, 193]. The last criticism doesn't occur in the German or, therefore, Woodger's translation; we can probably accept Hitchcock's conjecture [Stroińska and Hitchcock, 2002, 158] that Tarski added it to the Polish in response to discussion at the Paris congress. However, rather than softening the criticism of Carnap, as Hitchcock suggests, the second point emphasizes a perhaps more fundamental advantage of Tarski's account, which is the elimination of any dependence on transformation rules taken as primitively valid. This is the really revolutionary aspect of [Tarski, 2002].

Tarski then states a criterion of intuitive adequacy[2] for the definition of consequence:

> The point of departure for us will be certain considerations of an intuitive nature. Let us consider an arbitrary class of sentences $\mathfrak{K}$ and an arbitrary sentence X which follows from the sentences of this class. From the point of view of everyday intuitions it is clear that it cannot happen that all the sentences of the class $\mathfrak{K}$ would be true but at the same time the sentence X would be false. Since moreover it is a question here of the relation of following logically, i.e. formally, and therefore of a relation which has to be completely determined by the form of the sentences among which it obtains, thus following cannot depend on out knowledge of the external world, in particular on our knowledge of the objects which are spoken about in the sentences of the class $\mathfrak{K}$ or in the sentence X, cannot be lost as a result of our replacing the names of these objects in the sentences under consideration by names of other objects. Both these circumstances, which seem highly characteristic and essential for the concept of following, find jointly their expression in the following condition:
>
> *(F) If in the sentences of the class $\mathfrak{K}$ and in the sentence X we replace the constant terms which are not general–logical terms correspondingly by arbitrary other constant terms (where we replace equiform constants everywhere by equiform constants) and in this way we obtain a new class of sentences $\mathfrak{K}'$ and a new sentence X', then the sentence X' must be true if only all sentences of the class $\mathfrak{K}'$ are true* [Tarski, 2002, 183–4].

Tarski then notes that the condition isn't sufficient, since it may be satisfied only because some objects lack denoting constants in the language under consideration. One wouldn't want the inference from Fa to Ga to come out logical merely because some object that is *F* but not *G* lacks a name.

This, too, looks to be part of Intuitionistic Formalism: we have a consequence that is supposed to follow from a definition, by which the definition is to be judged. But condition (F) plays a different role from the T-sentences. The latter are such that implication of them by a deductive theory is sufficient for intuitive adequacy, while condition (F) is explicitly marked as necessary but insufficient. This is part of a shift of emphasis in the article: the definition of consequence, unlike definitions in earlier papers, appears to carry the weight of conceptual analysis itself. It isn't that adding the definition brings intuitive consequences; rather, intuitive consequences are evidence that a definition that is itself intended to be analytic of the ordinary concept is in fact such.

The solution to the problem that inferences may be accounted "logical" by condition (F) only because of the impoverishment of the object language is to replace substitution with semantics. Tarski explains the notion of satisfaction to the reader and defines a model of a set of sentences as a sequence that satisfies all of them. He then defines logical consequence in terms of models:

> We say that the sentence X follows logically from the sentences of the class
> ℜ if and only if every model of the class ℜ is at the same time a model of
> the sentence X [Tarski, 2002, 186].

Tarski then argues, in a passage subject to much dispute, that following according to condition *F* is a necessary condition of following according to Tarski's definition: the consequences of true sentences must be true, and arguments declared valid are all such that every one of their substitution instances has either a false premise or a true conclusion.[3] He then notes that his definition is essentially equivalent to Carnap's. Tarski ends the article by noting that the definition of consequence depends on the distinction between logical and non-logical constants and raises the possibility that this distinction cannot be drawn in a principled way, something he takes to be significant for Tractarian and Vienna Circle doctrines about the way in which logic "says nothing about the real world" [Tarski, 2002, 189].

Though I emphasize Tarski's attention to Carnap because it is Carnap that Tarski mentions, we should note that a definition of consequence resembling Tarski's was known in the Lvov–Warsaw School. Łukasiewicz gives this definition:

> The relation of implication, or the relation between reason and consequence,
> holds between two indefinite propositions a and b if for every pair of values of

the variables occurring in a and b either the reason a yields a false judgment or the consequence b yields a true judgement [Łukasiewicz, 1970a, 17].

This only concerns one-premised inferences and clearly the appeal to formality is lacking, but the similarity to Condition (F) is manifest. In something of an anticipation of notorious aspects of Tarski's view, Łukasiewicz happily notes that given this conception "x is greater than 4" has "x is greater than 3" as a consequence. In another article Łukasiewicz allows multiple premises and speaks of logic as involving "formal" consequence, but only in a degenerate sense of "formal" on which it amounts to containing variables:

> When we speak about consequence in logic, we usually mean *formal* consequence. The expression: a proposition is "a formal consequence" of another proposition, is used to denote a certain relation which holds between indefinite propositions or propositional functions. These two terms are used with reference to propositions which contain some variables ... Hence if we want to examine whether certain *definite* propositions are, or are not, formal consequences of other propositions, we must first transform all these propositions into indefinite propositions, that is, we must replace definite terms by variables represented by certain letters.
>
> The indefinite proposition Z is a consequence of the indefinite propositions $P, R, S, ...$, if there exist *no* values of the variables contained in the propositions $P, R, S, ..., Z$ which verify the propositions $P, R, S, ...$, but do *not* verify the proposition Z. [Łukasiewicz, 1970b, 66–7]

Notice that a feature of Tarski's own procedure, namely, replacing constants with variables rather than simply reinterpreting the constants themselves in the definition of logical consequence, is present here. This might be part of the explanation of Tarski's procedure, though of course given that Intuitionistic Formalism demanded that a constant get one and only one definite interpretation, Tarski would have found the move necessary anyway.

7.1.2 Objections to Tarski's account

The major known criticisms of Tarski's definition concern three basic, inter-related points about its adequacy and one question about Tarski's intentions. First, the effect of Tarski's definition is to treat the logical truth of a sentence as the plain truth of an associated universal generalization [Etchemendy, 1990, 95]. Etchemendy refers to this claim that logical truth reduces to the plain truth of the associated universal generalization on non-logical vocabulary as the "Reduction Principle" [Etchemendy, 1990, 98]. Especially given the fact that Tarski appears not to consider varying the domains of his models, the result is that a host of truths and falsehoods that are at the very best controversially "logical"—claims about how many objects there are, the continuum hypothesis, and so on—turn out to be "logical truths" or "logical falsehoods" on Tarski's definition. Similar points hold, *mutatis mutandis*, for logical consequence in general: logical consequence is simply material consequence among the associated generalizations; one sentence is a logical consequence of others iff either one of the generalizations of these others is false or its generalization is true. Arguments formulated with only vocabulary singled out as logical are therefore declared valid as long as they have either a false premise or a true conclusion. According to the intuitions of many, however, "there is one thing, therefore there are two things" is a paradigm of invalidity, yet on a standard selection of logical constants it comes out as valid on Tarski's account granted only that the universe doesn't contain exactly one thing. Call this, following Etchemendy's terminology, *the overgeneration problem*.[4]

Second, by treating logical truth and consequence in terms of the plain truth of associated generalizations, Tarski's definition seems entirely to lose the intuitive idea that a logical consequence of a set of sentences "must be" [Tarski, 2002, 187] true if the members of the set are, that "it cannot happen" [Tarski, 2002, 183] that the premises of a logically valid argument are true but its conclusion is false. If it just turns out that the generalization associated with a sentence is true, the sentence will be counted as logically true by Tarski's definition; as Etchemendy writes, "this happens whenever the language, stripped of the meanings of its non-logical constants, remains relatively expressive, or if the world is relatively homogenous, or both" [Etchemendy, 2008, 271–2]. Call this *the modality problem*.

Third, Tarski's conception of consequence is supposed to respect the intuition that logical consequence is tied to the "forms" of sentences [Tarski, 2002, 183], but, as Tarski notes at some length, the assumed account of "form" rests on a division of expressions of the language into logical and non-logical, a division that Tarski himself allows may admit of no principled characterization [Tarski, 2002, 188–90]. This last problem on its own might appear no more significant than any other noteworthy but manageable indeterminacy in an analytic category, but the issue becomes significant because it is held by many commentators that the first two criticisms might be answered by a proper and principled account of what makes an expression logical. This is where Tarski's

1966 "What are Logical Notions?" becomes significant, and we will turn to it in due course. Call this *the formality problem*.

Finally, turning to the question about Tarski's intentions, given the apparent difficulties of defending Tarski's definition of logical consequence as something like an analysis of what is often meant when one speaks in an ordinary way of valid arguments, there is a significant question in the literature as to whose conception of consequence Tarski was trying to represent, and to what standards of accuracy he held himself. Some commentators take it to be obvious that Tarski held himself responsible to the "ordinary" meanings of these terms, while others take it to be equally obvious that Tarski was attempting merely to define the notion of consequence as it appeared in mathematics, mathematical logic or "formal axiomatics" [Jané, 2006]. Call this *the analytic problem*. We will discuss it in the last section of this chapter, where we will evaluate Tarski's account.

7.2   Consequence in *Logical Syntax*

Although all of these problems are well worked over in the literature, surprisingly little is made of Tarski's many references to Carnap's *Logical Syntax*. Etchemendy, for instance, doesn't mention [Carnap, 2002] as a precursor to Tarski's analysis at [Etchemendy, 1990, 7], and other work has largely followed Etchemendy in the omission, [Coffa, 1987] and [de Rouilhan, 2009] being two noteworthy exceptions.[5] This is especially strange given that the only example of a definition of consequence Tarski mentions in the article is Carnap's, and that in the last paragraphs of the article Tarski addresses the significance of the distinction between logical and extra-logical terms for some typical Vienna Circle views.

7.2.1   L-consequence and condition F

Carnap's account of consequence is an improvement over a straightforward definition in terms of transformation rules of the sort that Tarski himself favored through 1934, and Tarski saw from *Logical Syntax* that it was headed in the right direction. He was also able to see, though, as we discussed in the previous chapter, that it breaks down at just the point where semantics can be used to fix it. The advance in Carnap's account was that by adding the "indefinite" rules of transformation, in particular the ω-rule, Carnap had a way of responding to Gödel's incompleteness theorem, on which derivability was insufficient to capture all sentences that are true in the language of a theory if the theory is true. Since Carnap eschewed "true" in favor of "analytic", he needed a way to state Gödel's result without the forbidden term, and introducing an infinitary transformation rule allowed him to state Gödel's result as the claim that there are consequences of theories extending arithmetic that aren't derivable.

The study of the "formal properties" of expressions proceeds in terms of a division of expressions into "syntactic" categories, which group symbols into types. In the preliminary discussions of Languages I and II, Carnap simply specifies the categories; for Language I we have variables, numerals, predicates, functors and, importantly for our purposes, eleven "individual symbols": the parentheses and comma, the successor stroke (which, see below, is not treated as a functor), four sentential connectives, the identity sign, and the signs "∃" and "K", which form parts of the existential quantifier and the K-operator—basically a kind of description operator we can gloss as "the least x less than y such that". In "General Syntax", by contrast, the syntactic categories are defined in terms of substitutions that preserve sentencehood: two expressions belong to the same "genus" just in case substituting one for the other in any sentence always produces a sentence, syntactic rules of formation for sentences being assumed [Carnap, 2002, 169]. An expression is generically "isolated" if it has a genus to itself.

In the treatment of Languages I and II the distinction between logical and descriptive vocabulary is made enumeratively: for Language I, any undefined predicate or functor, or any defined predicate or functor which is defined ultimately in terms of an undefined predicate or functor, is descriptive; all other symbols (including the successor stroke and the numerals) are logical.[6] An expression is descriptive if it contains a descriptive symbol, and logical otherwise. In general syntax, the definition comes in terms of consequence (which we will address in a moment): an expression is logical if and only if it belongs to the smallest set of expressions such that every sentence formed out of them is a consequence of the empty set of sentences or has every sentence as a consequence [Carnap, 2002, 177–178]. Since any sentence containing a primitive predicate or functor of Languages I and II (or something defined in terms of one) is "indeterminate"—neither a consequence of the empty set of sentences nor having every sentence as a consequence— the characterization matches the treatment of descriptive expressions in the treatment of languages I and II (the successor stroke, which officially wasn't a "functor" in I and II ends up being logical because, given the transformation rules stated for I and II it belongs to the set of expressions such that every sentence formed only from them is determinate, despite the fact that sentencehood is preserved by substituting various functors for it). A language is logical just in case all of its sentences are determinate.

Logicality, then, in general syntax, assumes a primitive notion of consequence. Two distinctions are relevant here; Carnap has direct and indirect consequence and derivability. A sentence is directly derivable from others, or is a direct consequence of them, if and only if there is a defininte transformation rule that states as much [Carnap, 2002, 168]; repeated applications of these rules give rise to indirect derivation [Carnap, 2002, 171–2]. The distinction between consequence and derivability has to do with transformation rules that are "definite" as opposed to "indefinite". This distinction, in general, concerns whether or not unrestricted quantification over infinite totalities is involved; Carnap's Language I is definite in that its quantifiers are always finitely restricted. When it comes to transformation rules, these are indefinite when they state that a certain sentence is a transformation of some infinite set of sentences. The example in the text is the ω-rule, which we have seen Tarski call "the rule of infinite induction" [Tarski, 1983l, 259], which appears as Carnap's DC2 in the discussion of Language I [Carnap, 2002, 38]. As Carnap notes, although in all but the simplest languages it is not the case that every sentence formed with only logical expressions is either derivable or refutable, with the ω-rule added, every such sentence or its negation is either consequence of the empty set of sentences or has every sentence as a consequence [Carnap, 2002, 40]. If only Carnap allowed himself to make serious use of the notion of truth, he would be able to say that analytic sentences are those that are true in virtue of their syntactic form, but as it is he has to relegate the point to an informal gloss:

> By means of the concept "analytic", an exact understanding of what is usually designated as "logically valid" or "true on logical grounds" is achieved. Hitherto it has for the most part been thought that logical validity was representable by the term "demonstrable"—that is to say, by a process of derivation. But although, for the majority of practical cases, the term "demonstrable" constitutes an adequate approximation, it does not exhaust the concept of logical validity. The same thing holds for the pairs "demonstrable"–"analytic" and "refutable"–"contradictory", as for the pairs "derivable" "consequence".
>
> In material interpretation, an analytic sentence is absolutely true whatever the empirical facts may be. Hence, it does not state anything about facts ... A synthetic sentence is sometimes true—namely, when certain facts exist—and sometimes false; hence it says something as

to what facts exist. Synthetic sentences are the genuine statements about reality [Carnap, 2002, 41].

This is, however, simply "material interpretation"; the official position is simply that an analytic sentence is a consequence of the empty set, and consequence outstrips derivability; the informal notion of logical truth is formally taken up by being a consequence of everything. Note the work that the notion of analyticity here does in upholding Carnap's *Tractatus*-inspired conception of logic and the correlative treatment of mathematics. We will return to this in a moment.

We come finally to Carnap's understanding of "L-consequence", which Tarski takes as the notion of logical consequence and defines in what he believes to be a more complete and suitable manner. Carnap informally distinguishes "L-rules" from "P-rules" in terms of the former's being "logico-mathematical" [Carnap, 2002, 180] and writes:

Now how is the difference between L-rules and P-rules—which we have here only indicated in an informal way—to be formally defined? This difference, when related to primitive sentences, does not coincide with the difference between logical and descriptive sentences. An $\mathfrak{S}_l$ as a primitive sentence is always an L-rule; but an $\mathfrak{S}_{\mathfrak{d}}$ as a primitive sentence need not be a P-rule. [Example: Let 'Q' be a $\mathfrak{pr}_{\mathfrak{d}}$ of Language I. Then, for example,

$$\text{'Q}(3) \supset (\sim Q(3) \supset Q(5))\text{'} \ (\mathfrak{S}_1)$$

is a descriptive primitive sentence of the kind PSI i. But $\mathfrak{S}_1$ is obviously true in a purely logical way, and we must arrange the further definitions so that $\mathfrak{S}_1$ is counted amongst the L-rules and is called, not P-valid, but analytic (L-valid). That $\mathfrak{S}_1$ is logically true is shown formally by the fact that every sentence which results from $\mathfrak{S}_1$ when 'Q' is replaced by any other $\mathfrak{pr}$ is likewise a primitive sentence of the kind PSI i.] The example makes it clear that we must take the general replacability of the $\mathfrak{A}_{\mathfrak{d}}$ as the definitive characteristic of the L-rules [Carnap, 2002, 180–1].

Carnap here expresses his view about what it takes to be "true in a purely logical way": a sentence is analytic if it follows from the empty set and continues to do so under all uniform substitutions of non-logical vocabulary. Working this idea out gives us what we can recognize as the basic idea Tarski works with in condition (F): a sentence is an L-consequence of a set of sentences just in case every uniform substitution of non-logical expressions produces a sentence and a set such that the sentence is a consequence of the set [Carnap, 2002, 181]. Here is Carnap's definition; remember that consequence itself is simply a matter of stipulated transformation rules and that the definition sorts consequences into L-consequences and the rest:

> Let $\mathfrak{S}_2$ be a consequence of $\mathfrak{K}_1$ in S. Here three cases are to be distinguished. 1. $\mathfrak{K}_1$ and $\mathfrak{S}_2$ are logical. 2. Descriptive expressions occur in $\mathfrak{K}_1$ and $\mathfrak{S}_2$, but only as undefined symbols; here two further cases are to be distinguished: 2*a*. for any $\mathfrak{K}_3$ and $\mathfrak{S}_4$ which are formed from $\mathfrak{K}_1$ (or $\mathfrak{S}_2$) by the replacement of every descriptive symbol of $\mathfrak{K}_1$ (or $\mathfrak{S}_2$ respectively) by an expression of the same genus, and specifically of equal symbols by equal expressions, the following is true: $\mathfrak{S}_4$ is a consequence of $\mathfrak{K}_3$ ...

This would be complete if it weren't for the fact that the use of defined symbols can disguise the underlying form of the sentence, and in particular whether a defined predicate is ultimately defined in terms of logical or non-logical primitives [Bays, 2001, 1724]; hence we need to consider the third case:

> 3. In $\mathfrak{K}_1$ and $\mathfrak{S}_2$ defined descriptive symbols also occur; let $\bar{\mathfrak{K}}_1$ and $\bar{\mathfrak{S}}_2$ be constructed from $\mathfrak{K}_1$ (or $\mathfrak{S}_2$ respectively) by the elimination of every defined descriptive symbol (including those which are newly introduced by the result of an elimination); 3*a*. the condition given in 2*a* for $\mathfrak{K}_1$ and $\mathfrak{S}_2$ is fulfilled for $\bar{\mathfrak{K}}_1$ and $\bar{\mathfrak{S}}_2$ ... In cases 1, 2*a*, 3*a*, we call $\mathfrak{S}_2$ an **L-consequence** of $\mathfrak{K}_1$ ... Thus the formal distinction between L- and P-rules is achieved [Carnap, 2002, 181].

Tarski's condition (F) is simply a modification that replaces the appeal to consequence defined in terms of primitive transformation rules with the idea that if the premise-set is true, the conclusion must be true.[7] (Tarski leaves out the third condition only "for the sake of simplifying our considerations" [Tarski, 2002, 184]). Tarski's aim is simply to restore the intuitive connection, broken by Carnap's skepticism about semantics, between logical consequence and truth, and to make up for the definition's reliance on substitution—the two aspects of Carnap's skepticism about semantics we discussed in § 3.1.

7.2.2 Tractarianism in the Vienna circle

That, then, is the basic idea: Tarski is simply out to adjust Carnap's definition of L-consequence from General Syntax by replacing a substitutional interpretation of L-validity with an objectual one, and transformation rules with a semantic account of consequence that implies that the consequences of true sentences are true. However, Tarski also had an axe to grind with Carnap. Carnap, along with the majority of the Vienna Circle, held that the Tractarian doctrine that logic says nothing about the world is crucial to a scientific worldview, since, on their view, it was the only satisfactory conception of mathematics compatible with empiricism:

> The conception of the nature of mathematics which we developed in the discussions of the Vienna Circle came chiefly from the following sources. I had learned from Frege that all mathematical concepts can be defined on the basis of the concepts of logic and that the theorems of mathematics can be deduced from the principles of logic. Thus the truths of mathematics are analytic in the general sense of truth based on logic alone. The mathematician Hans Hahn, one of the leading members of the Circle, had accepted the same conception under the influence of Whitehead and Russell's work, *Principia Mathematica*. Furthermore, Schlick, in his book *Allgemeine Erkenntnislehre* (1918), had clarified and emphasized the view that logical deduction cannot lead to new knowledge but only to an explication or transformation of the knowledge contained in the premises. Wittgenstein formulated this view in the more radical form that all logical truths are tautological, that is, that they hold necessarily in every possible case, therefore do not exclude any case, and do not say anything about the facts of the world. Wittgenstein demonstrated this thesis for molecular sentences (i.e. those without variables) and for those with individual variables. It was not clear whether he thought that the logically valid sentences with variables of higher levels, e.g., variables for classes, for classes of classes, etc., have the same tautological character. At any rate he did not count the theorems of arithmetic, algebra, etc., among the tautologies. But to the members of the Circle there did not seem to be a fundamental difference between elementary logic and higher logic, including mathematics. Thus we arrived at the conception that all valid statements of mathematics are analytic in the specific sense that they hold in all possible cases and therefore do not have any factual content.

What was important in this conception from our point of view was the fact that it became possible for the first time to combine the basic tenet of empiricism with a satisfactory explanation of the nature of logic and mathematics. Previously, philosophers had only seen two alternative positions: either a non-empiricist conception, according to which knowledge in mathematics is based on pure intuition or pure reason, or the view held, e.g., by John Stuart Mill, that the theorems of logic and of mathematics are just as much of an empirical nature as knowledge about observed events, a view which, although it preserved empiricism, was certainly unsatisfactory [Carnap, 1963, 46–7].

Carnap saw the treatment of logic and mathematics in *Logical Syntax*, on which "In material interpretation, an analytic sentence is absolutely true whatever the empirical facts may be. Hence, it does not state anything about facts" [Carnap, 2002, 41] as a way of continuing to work out the basic logicist reconciliation of empiricism with the status of mathematics:

> The idea occurred to me that … there seemed to be a possibility of reconciling the conflict between intuitionism and formalism. Suppose that mathematics is first constructed as a purely formal system in Hilbert's way, and that rules are then added for the application of the mathematical symbols and sentences in physics, and for the use of mathematical theorems for deductions within the language of physics. Then, it seemed to me, these latter rules must implicitly give an interpretation of mathematics. I was convinced that this interpretation would essentially agree with the logicist interpretation of Frege and Russell [Carnap, 1963, 48].

Since, as *Logical Syntax* held, "*In logic, there are no morals*" [Carnap, 2002, 52], Carnap felt free to stipulate formation and transformation rules that had the consequence that mathematical statements were "analytic" in the sense of being derivable from everything. Such languages were intended to be applied in science in accord with coordinating definitions [Carnap, 2002, 78], but the question of the status of mathematics, and its reducibility to logic, were avoided *via* the tolerant conception of logic. (On the development from Frege's logicism and the role of such coordination, see [Ricketts, 2004] and [Ricketts, 2009].) The cornerstone of the approach, however, was still the rigid distinction between logic and factual statement which, combined with logicism and the Tractarian idea that since logical truths exclude nothing they aren't really "about reality", obviated the perennial questions about the status of mathematical objects and our knowledge of them. As I will argue below, Tarski's remarks on the logical constants, and his claim that logical consequence degenerates to material consequence if all expressions of the language are taken to be logical, are intended to express his disagreement with these views.

7.3   The overgeneration problem and domain variation

The interpretive hypothesis is that Tarski, in his treatment of consequence itself, intends in [Tarski, 2002] only to modify Carnap's definition
of L-consequence by supplying, via semantics, objectual quantification
in favor of Carnap's substitutional quantification, and truth-preservation
in favor of primitive rules of "direct consequence". The evidence for the
hypothesis is of several sorts. First, that Tarski presented the paper to a
room full of Carnap and his friends and colleagues and the Paris Unity
of Science congress. Tarski would have wanted to showcase what he had
to offer while arousing as few extraneous and distracting disputes as possible. He therefore assumed all elements of Carnap's conception save
those he specifically wanted to replace with semantics. Second, Tarski's
adequacy condition is an obvious modification of Carnap's definition of
L-consequence. Finally, the remaining features of the article—namely,
the skepticism about the distinction between logical and non-logical
constants the definition itself uses—are to be explained by Tarski's disagreement with Carnap over the Tractarian conception of logic. The best
test of the hypothesis, however, is in the treatment of the standard interpretive puzzles. Let us now see how the proposed reading fares with
the overgeneration, modality, formality, and analytic problems, starting
with the first.[8]

7.3.1   Domain variation

The essence of the overgeneration problem is that, by treating logical
truth as the plain truth of the universal generalization of a sentence
with respect to its non-logical vocabulary, Tarski's definition declares
any sentence formulated only in logical vocabulary as logically true or
logically false. Yet many such sentences, such as $(\exists x)(\exists y)(x \neq y)$, seem
neither logically true nor logically false. The standard model-theoretic
account of later decades avoids this consequence by varying the domain
of quantification across interpretations so that this sentence is neither
logically true nor logically false, as there are interpretations with domains
of fewer than two elements. Sympathetic interpreters here fall into two
categories, maintaining either that Tarski's account in fact doesn't have
the consequences deemed objectionable because Tarski actually assumed
such variation as read, or that these consequences were not in fact objectionable to Tarski because he didn't think that logic should lack such
mathematical content.

The first account takes its inspiration from Tarski's discussion of the
concepts of "*satisfaction* and *correct sentence ... with respect to a given*

individual domain a" at, e.g., [Tarski, 1983a, 239–40] (§ 4.2.3). Since in this passage Tarski explicitly recognizes the possibility of relativizing the domain of quantification and in particular of the possibility of studying which sentences remain "correct" in every relativized domain, and since he notes that "the point of view is better established" that logic is only concerned with sentences correct in every individual domain, there is plausibility in the suggestion that Tarski assumes in [Tarski, 2002] that the domain varies across interpretations, and this suggestion is the basis of the family of interpretations on which it is argued that Tarski's account isn't subject to the overgeneration problem because Tarski assumes without mentioning it that the domain of quantification is supposed to vary across interpretations, e.g. [Ray, 1996, 628–30], [Sher, 1991, 45ff], [Gómez-Torrente, 1996, 143–5].[9] The well-known weakness of the reading is that Tarski nowhere in [Tarski, 2002] says that the domain varies and, moreover, that Tarski holds that logical consequence reduces to material consequence when all expressions of the language are treated as "logical", which would be wrong if the domain varied, since "there is one thing, therefore there are two things" would have a true premise and a false conclusion in a one-element domain.

Attempts to read domain variation into [Tarski, 2002] take two forms. One either (i) argues that Tarski just is assuming that the sequences that form his models are of variable length without mentioning it ([Sher, 1996], [Ray, 1996] (endorsed by [Hanson, 1999, 610] and Stroinska and Hitchcock's introduction to [Tarski, 2002, 170]), or (ii) argues that though Tarski's sequences are all infinite, suitable use of implicit restriction of the range of the variables gives Tarski the effect of variable domains within the confines of the conception more clearly on the surface of [Tarski, 2002] ([Gómez-Torrente, 1996]). Our way with (i) can be brief: as Etchemendy notes [Etchemendy, 2008, 280], there is simply no evidence that Tarski assumes that the domain varies in [Tarski, 2002] and the reading is incompatible with Tarski's claim that logical consequence degenerates to material consequence if all terms are treated as logical. Ray holds that this is simply a mistake on Tarski's part, but by his own admission this depends on the assumption that it is a mark of greater stupidity to think that that "there are two things" is a logical truth than it is to think that "there are two things" is true in a one element domain. Like most interpreters [Bays, 2001, 1709], [Etchemendy, 2008, 280], [Gómez-Torrente, 2009, 268], I find this implausible.

The stronger approach is (ii). Gómez-Torrente pushes the line that Tarski often formulated theories with a predicate D characterizing the domain of objects with which the theory was concerned

[Gómez-Torrente, 1996, 143]. The idea here is that since this predicate is non-logical, it will receive different interpretations in different models, and since it specifies what the theory is about, the variables with which other non-logical constants are replaced in Tarski's test for consequence will range over it and hence over a variable domain. Thus $(\exists x)(\exists y)(x \neq y)$ is really shorthand for $(\exists x)(\exists y)((Dx \ \& \ Dy) \ \& \ (x \neq y))$ and, the idea is, this is false on any Tarskian model where the variable that replaces D is assigned an extension of < 2 elements and the sentence is thereby not classified as logically true.

However, as Mancosu points out [Mancosu, 2006, 220–6] the discussion of categoricity in [Tarski, 1983k] and other passages make clear that Tarski, following Langford, allowed an indeed often demanded that theories quantify over objects not within the extension of D. Since Tarski is working within STT, this means that for the theories he considers in general $D \neq V$ and thus that the variability of the extension of "D" implies nothing about the variability of V and, correlatively, that the quantifiers aren't to be read as implicitly relativized to D. Note also that Mancosu distinguishes "strong" from "weak" fixed domain views, where a strong view holds that V is fixed once and for all in every application of STT (presumably to "all objects") while the weak view is simply that though the domain of the quantifiers is fixed to be V in any given definition of consequence for STT, V itself is taken to have different cardinalities in different applications. As Mancosu also points out, it was common for other authors, in particular Gödel, to work in STT with V fixed to be something other than everything that exists [Mancosu, 2006, 232–6]. All that is shown by Tarski's variation of V across examples is that V wasn't conceived of by him or others to be required actually to be the entirety of the universe *in a particular application.*[10]

In order to sort things out here we need to introduce a previously neglected distinction. One question is whether Tarski in the 1920s and 1930s had the idea of a set of sentences being satisfied by different sets of objects and relations upon them, where, in particular, these sets could vary in cardinality. A second question is whether in the same period Tarski made some use of the idea that models of various cardinalities were quantified over *in the definition of consequence itself.* It is beyond question that Tarski had the first idea from very early on [Feferman, 2008, 78–9]. What isn't clear at all is that such domain variation played any role in his conception of logical consequence and logical truth. The textual evidence marshalled by Gómez-Torrente establishes the first, but only the ill-supported suggestion that, in at least some applications *of the definition of consequence* Tarski ties the range of the variables to the extension of the domain predicate, makes it look as though the answer to the second is in the affirmative. The mere fact that Tarski had the idea that that some theories, or even logic itself, consisted of sentences that are true in domains of variable cardinality doesn't of itself establish that he conceived of semantic consequence as he introduced it in terms of models with domains of variable cardinality.

This distinction in hand, we can consider the fact that [Gómez-Torrente, 2009, 255ff] responds to Mancosu with many examples, mostly taken from the textbook, of Tarski taking it that a change in the extension of the domain predicate of a theory correlates with a change in the intended range of the variables in its sentences. The problem is that the examples are all easily accounted for by Mancosu's idea [Mancosu, 2006] that Tarski allows V to vary across applications but doesn't hold that $D = V$, as long as we assume, as is reasonable, that if there is any place that Tarski *isn't* being fully explicit about the logical backdrop of the theories he considers, it is in the toy examples of the textbook. The passages Gómez-Torrente cites never say that the range of the variables changes *because* the domain predicate has a different extension; they simply rather inarticulately make the novice reader familiar with the idea of varying the interpretation of a theory in an intuitive way. I can't respond to all points of detail here, but for the sort of thing I have in mind, consider Gómez-Torrente's argument that on the fixed domain conception of the range of the variables we get the result that contradictions are provable in the theory if the extension of the domain predicate is moved outside of the fixed range of the variables [Gómez-Torrente, 2009, 258]. This, however, assumes the "strong" fixed domain conception; the "weak" conception, by contrast, handles it just fine: when Tarski considers reinterpreting his toy theories in the textbook, he changes the interpretations of both the range of the variables and the domain predicate, without assuming that the range of the variables is settled by the extension of the predicate.

At most establishing that Tarski contemplated theories in which, e.g. $(\exists x)(\exists y)(x \neq y)$ is really shorthand for $(\exists x)(\exists y)((Dx \& Dy) \& (x \neq y))$ would establish that Tarski's account of logical consequence gives reasonable verdicts on what follows logically from non-logical theories formulated in this manner. It wouldn't address what Etchemendy takes to be the pressing problem, namely, counting $(\exists x)(\exists y)(x \neq y)$ as a plain logical truth. It is true that Tarski can get the effect of variable domains for a theory stated within STT through restricting everything by a domain predicate, but this does not show that Tarski allows that the notion of logical consequence for STT itself is a variable-domain notion.

A potential fly in the ointment here is Tarski's discovery of the upward Löwenheim–Skolem theorem in the late 1920s, as announced in a note in *Fundamenta Mathematicae* [Mancosu et al., 2009, 433–4]. After all, if the result is that any first-order theory with a countably infinite model has a model in every infinite cardinality, isn't Tarski thinking of models as varying in cardinality? However, this problem is easily solved on either Gómez-Torrente's or Mancosu's approaches, since the result is that any *theory* expressed first-order that has a countable model has models in further infinite cardinalities. We have already seen that variable domains for a theory can easily be accommodated by Tarski's various approaches at the time, even if V is held fixed in STT and thereby in the definition of consequence. For the present application, simply let V be "everything" or at least some proper class, and let D be a non-logical domain predicate; the general Löwenheim–Skolem result is that a first-order theory can at most force the extension of D to be infinite, but cannot settle its cardinality any further than that. None of this goes to show that the range of the variables in the language of the theory itself changes across interpretations, and hence none of it goes to show that in the test for consequence itself Tarski is allowing the range of the variables to vary.

The position that emerges is coherent. Logic itself assumes a fixed domain: all individuals.[11] When defining logical consequence and logical truth this is the invariant domain (cf. [Milne, 1999, 158–9]). However, a matter of interest is to consider what follows logically from a theory stated within a background logic when *its* domain, specified by a non-logical predicate, varies in accord with a perfectly standard application of the (domain invariant) definition of consequence. When things are done this way, of course, a "variable domain" conception of what follows *from the axioms of that theory* arises automatically: the logical consequences of the axioms are just what follows from them by the fixed-domain account of consequence, in that the inclusion of the domain predicate forces what follows logically from the axioms to be what follows no matter what the cardinality of D's extension is. What follows logically from the theory is then automatically what follows independently of the cardinality of its domain. Seen in this light, Tarski's investigation of theories with domain predicates is perfectly compatible with a conception of logic proper that assumes a fixed domain of all individuals.

7.3.2   Consequence in Gödel's completeness theorem

The preceding point in hand, we can confront the suggestion that is sometimes made that Tarski was aware of variable domain conceptions of consequence because they played a role in the statement of the completeness theorem for first-order logic. Gómez-Torrente [Gómez-Torrente, 1996, 138] takes the fact that Tarski was aware passages like this one from Hilbert and Ackermann as evidence that he was familiar with variable domain conceptions of consequence:

> Whether our axiom system is complete (*vollständig*), at least in the sense that all logical formulae that are correct (*richtig*) in every domain of individuals (*Individuenbereich*) can actually be derived, is yet an unsolved question [Hilbert and Ackermann, 1928, 68] (my translation).

The weakness here, however, is as above: Tarski's familiarity with Gödel's completeness theorem shows only that he had the notion of truth in domains of all cardinalities, and not that his conception of consequence itself involved varying the cardinality of the domain. However, the passage simply does not express a variable domain conception of consequence. What Hilbert and Ackermann do is this. They have a language, and an axiomatic theory in that language. The set of sentences of the language has a certain subset, namely, the set of those sentences of the language that are "correct" (that is, true) in every individual domain. Now the question arises: can all of the sentences of that set be derived from the axioms by the rules of inference given? There is nothing here about these sentences bearing some relation other than derivability to the axioms, nothing, e.g. about there being a non-derivational relation of "following" that sentences can bear to the axioms only in a way that is independent of the cardinality of the domain. The question is simply whether those sentences correct in every individual domain are derivable from the axioms. The only conception of consequence at work in the question is derivability. The passage simply provides an example of what we know Tarski was aware of, that the notion of truth in a given, or every, individual domain was of interest and that the correlative questions about the relations between provability and correctness in every individual domain were of interest. The passage doesn't show that Hilbert and Ackermann had a non-derivational, semantic, variable domain conception of logical consequence in 1928 (cf. [Etchemendy, 1988, 75]).

Nor does [Gödel, 1967a] show that Gödel had such a conception. Gödel's conception of the question he answers is exactly Hilbert and Ackermann's: is every sentence of Hilbert and Ackermann's "restricted functional calculus" that is "valid in every individual domain" [Gödel, 1967a, 584] provable from the axioms? There is in Gödel's paper no more than in Hilbert and Ackermann's book any definition or implicit conception of "following from" in terms of models, be they of variable cardinality or not. The question is simply whether all sentences that are valid in every individual domain are derivable from the given axioms by the given rules.[12]

There is, in hindsight, a conceptual restriction here: why be interested merely in whether every sentence that is true in every individual domain is provable from the axioms by the rules, and not in the more general question of whether every sentence s and set Γ such that in every domain either a member of Γ isn't true or s is, is also such that s can be derived from Γ given the axioms and rules of inference? If the second question hadn't reduced to the first (via the deduction theorem) in the case Hilbert and Ackermann and Gödel considered, a model-theoretic conception of of consequence would have had to have been arrived at by Hilbert and Ackermann in order to formulate the question that Gödel then answered.[13] Given that their interest was precisely in "correctness" in every domain, such a conception of consequence would have involved variable domains and history in the relevant respects might have been different. As it was, however, the second question did reduce to the first and so no conception of consequence other than derivability was required to formulate the completeness theorem as Gödel proved it (cf. [Etchemendy, 1988, 68]). We should not let our familiarity with the now-standard statement of the theorem that if $\Gamma \models s$ then $\Gamma \vdash s$ make us overlook this (cf. [Etchemendy, 1988, 76-7].

Now there are hints of at least a distinction between derivability and consequence in some authors known to Tarski and predating Gödel. In Veblen's work, e.g. [Veblen, 1904, 346], as suggested by Mancosu, Zach and Badesa [Mancosu et al., 2009, 328] and [Veblen, 1906, 28], as suggested by Awodey and Reck [Awodey and Reck, 2002, 18] one finds at least a rudimentary distinction between semantic consequence (if the axioms are true, then a given sentence is true) and derivability. The Veblen passages are tantalizing and if Veblen intended to say what he appears to say he was well ahead of his time, but the suggestions are so underdeveloped that it would be difficult to ascribe any historical influence to them. More proximally, Fraenkel [Fraenkel, 1928, 347-9] clearly distinguishes derivability from semantic consequence and doubts that all semantic consequences of a theory must be derivable (according to [Awodey and Reck, 2002, 22], following Weyl). However, in neither Fraenkel nor in Veblen is there anything about models varying in cardinality or the interest being in semantic consequence in all cardinalities. The notion at work is the same fixed domain semantic notion of consequence with which we have seen Tarski was familiar from Łukasiewicz.

But it is the variable notion that the objection we are considering in this subsection claims must have been known to Tarski from earlier authors. Contrary to the objection, however, the situation is the same in Tarski's predececessors as it is in Tarski's work itself, [Etchemendy, 1988, 70]: sometimes people worked with variable domains, sometime people varied the interpretation of non-logical expressions, but nobody was putting the two ideas together into the model-theoretic conception of consequence as we have it today (cf. [Milne, 1999, 159]).

I mention this point because it goes to the heart of the matter when it comes to the variable domain readings of Tarski. The strongest point in favor of these readings is, supposedly, that Tarski and others, well before [Tarski, 2002], were thinking about the relation of truth or "correctness" that held between sentences and domains of variable cardinality. On this basis, these readings then suggest that since Tarski was familiar with variable domain conceptions *of models* he must have intended to offer a variable domain conception *of logical consequence* in [Tarski, 2002], or even that a variable domain conception of consequence is "presupposed" by the completeness theorem, which Tarski obviously knew and understood. The problem with this suggestion is that what we find at [Tarski, 1983a, 239–40], [Hilbert and Ackermann, 1928, 68] or [Gödel, 1967a] isn't anything that makes any connection at all between the idea of a set of sentences having models (of variable cardinality or not) and the concept of logical consequence itself. We simply find the idea that there are sentences that are true regardless of the cardinality of the domain and then the question of whether these sentences are *derivable* in this or that system. Of course, especially in hindsight, the move from thinking of sentences that are true in every domain to thinking of sentences that are, in every domain, true if certain others are true, is not a big step. But what was a big step was the move from thinking of consequence in terms of the repeated application of a given, simple set of primitive forms of inference to thinking of it as having anything to do with the relation between sentences and their models. Only the revelation of Gödel's *in*completeness theorem could undermine the standing faith in the idea that logical consequence was simply a matter of derivability via repeated application of certain obviously valid rules; the completeness theorem alone had precisely the opposite effect.

7.3.3 Tarski's fixed domain

A further point needs to be made, in light of this discussion of Gödel, about Tarski's relationship to Carnap. Carnap took seriously the idea that what Gödel showed in the first incompleteness theorem was, at least in part, that the idea of one sentence following from others couldn't be cashed out in terms of derivation. Carnap drew the conclusion that infinitary "rules of proof" could close the gap, and indeed they could, but reliance on them stretched the meaning of "proof" too far. Tarski recognized that Carnap's treatment of L-consequence took an essential step away from a derivational conception of consequence, but also that, due to Carnap's Tractarian hostility to talk of word–world relations and his concomitant reliance on substitutional interpretations of generality, Carnap's notion was still too tied to the actual expressive power of the languages to which it was applied, and that it was hobbled by its reference to the assumed transformation rules. Working from Carnap, he replaced the reliance on substitution in Carnap's conception of L-consequence with the notion of satisfaction, and the result, to that point, was the definition of logical consequence as we have it in [Tarski, 2002]. Conceiving of consequence in terms of models at all was, then, a significant conceptual leap; as noted above, interpretive suggestions to the contrary notwithstanding, the idea simply isn't to be found in commonly cited passages of Hilbert and Ackermann or Gödel. Tarski then held to the fixed domain conception of logical consequence at least through 1940 [Mancosu, 2006]. Nothing in the treatment of truth in an individual domain in his writings or the writings of those he mentions before [Tarski, 2002] speaks against this.

With consequence defined as he defined it, Tarski was still free to consider which sentences of various theories were true in various individual domains, and to see which followed logically from these theories when they were restricted by a non-logical domain predicate the extension of which therefore varied, but all of this took place against a background conception of logical consequence itself in which the domain was fixed. Now, since Tarski was at most but a stone's throw from applying a variable domain conception of models in the definition of consequence in 1936, we have to ask why he operates resolutely with a fixed domain of quantification. Moreover, we also have to address a question about Gödel's completeness theorem as Tarski knew it and his definition of consequence: as Etchemendy points out [Etchemendy, 1988, 70], since Gödel properly understood did treat logical *truth* as truth in domains of all cardinalities, Tarski has to have known that his definition in [Tarski, 2002] implied that Gödel hadn't shown that first order logic is actually "complete", since Tarski's models have a fixed infinite domain and the deductive system Gödel considered couldn't prove the truths about this domain that Tarski's definition declares logical. The question in the body of this section was whether the models quantified over in the definition of consequence itself are of varying cardinality. The question here is the subsidiary historical one of why Tarski doesn't until surprisingly late combine two ideas that he had had in the 1930s to produce the definition of consequence that became standard (and accepted by him) by the 1950s.

Here I think two relatively standard suggestions are essentially right. The first is that since Tarski took STT to be logic, he simply didn't think the account "overgenerated" in Etchemendy's sense.[14] Tarski never explicitly talks about logicism (see [Murawski, 2004] on the general attitude toward such questions in Warsaw), but he was certainly happy to countenance higher-order logic as logic, and with this came, via the usual definitions, mathematics. The second standard suggestion (e.g. [Milne, 1999]) is also important. In the 1930s Tarski's target notions are conceived of as absolute: being true, and being true if other things are, in the universe as it is. Tarski took the question of what followed logically from what to be the question of what was true under all interpretations, holding logical vocabulary fixed, *period*, not what was true in this or that subregion of the universe, or what was true in every subregion of the universe. Logic was, for Tarski at the time, about what was true given that other things were, no matter how non-logical vocabulary was understood. Truth in an, or every, individual domain was an interesting thing to consider, too, but it had nothing to do with the relation of logical consequence.[15]

7.4 The modality problem and "Tarski's Fallacy"

A virtue of the interpretive approach based on tying [Tarski, 1983h] to [Carnap, 2002] is that it gives us a simple and natural reading of a passage in the article that has been the subject of a great deal of contention: Tarski's claim, immediately upon stating the definition of logical consequence, that "it can be proved on this basis that the consequences of true sentences must be true" [Tarski, 1983h, 417]. The literature here consists of a debate over how Tarski could possibly have thought that the definition allowed him to "prove" that the relation defined had any modal feature, with Etchemendy suggesting that Tarski is simply caught up in a transparent modal fallacy and rival commentators either suggesting that Tarski could establish that his conception of consequence does have modal import of some kind, downplaying the conception of this import so as to make the supposed proof easier, or some combination of the two.

7.4.1   Modalities

On the first side we have Etchemendy, who on my view quite rightly points out that:

> The most important feature of logical consequence, as we ordinarily understand it, is a modal relation that holds between implying sentences and sentence implied. The premises of a logically valid argument cannot be true if the conclusion is false; such conclusions are said to "follow necessarily" from their premises.
>
> That this is the single most prominent feature of the consequence relation, or at any rate of our ordinary understanding of that relation, is clear from even the most cursory survey of texts on the subject (1990, 81).

Etchemendy is also right that Tarski's view supports nothing of the sort. As he puts it at [Etchemendy, 2008, 267], if the validity of an argument comes down to nothing more than the material validity of its variants under uniform substitution of non-logical vocabulary (or, more generally, Tarski's treatment of this in terms of assignments to substituted variables), then we have no alethic modality, since it may simply be accidentally true that there are no such variants with true premises and a false conclusion, and we have no epistemic modality either, since when presented with an argument whose premises are known to be true, all that one could conclude is that either its conclusion is true or the argument is, by Tarski's definition, invalid [Etchemendy, 2008, 267].

As noted above, the responses to Etchemendy on this point either insist that somehow his account can support some modality, or insist that "must" is intended weakly enough that Tarski's claim to be able to establish that the consequences of true sentences "must be" true is acceptable. Since it is amply clear that Tarski's account involves the "Reduction Principle" on which logical truth is simply the plain truth of the universal generalization of a sentence on its non-logical vocabulary, the only recourse for an interpreter who wants to read a significant alethic modality into Tarski's account is the expedient of taking the account to quantify over non-existent objects. Commentators who argue forthrightly for this are hard to find these days, though Popper, always an idiosyncratic interpreter of Tarski (cf. § 4.3.4) in 1959 conflates Tarski's models with possible worlds at [Popper, 2002, 432] while arguing that Tarski did successfully reduce the notion of "logical necessity" to generality on the way to proposing his own parallel account of "natural necessity". A number of contemporary interpreters do make at least tentative suggestions in this direction, usually as part of a defense of Tarski on the overgeneration problem on which it is claimed that the account's implying that the existence of *n* things is a logical truth for many values of "*n*" is acceptable. The strategy tempts, since if the quantifiers in the generalization of a sentence quantify over mere *possibilia*, overgeneration seems less problematic—it being less of a stretch to think that logic tells us how many things might exist than how many things do exist. A modal aspect is thereby also secured for the consequence relation, since an argument will be declared logically valid not merely if all relevant sets of actual objects that do satisfy the logical forms of the premises also satisfy that of the conclusion, but if all relevant sets of possible objects do.

Priest, for instance, suggests that Tarski's account doesn't overgenerate because the quantifiers should be taken to range over mere *possibilia* and that there are as many of these as one pleases "isn't a substantive fact at all" [Priest, 1995, 291]. He further suggests that this solution to the problem of overgeneration is acceptable because "the modality involved seems just what is required if Tarski's reduction of a modal notion to a quantifier is to be right" [Priest, 1995, 291]. Sher [Sher, 1996, 681–2] likewise argues that logic on her "Tarskian" account is concerned with "formal possibility" and suggests that mathematics is the theory of formal possibility, though it is less clear than with Priest that she attributes the view to Tarski himself. As with Priest's account, taking mathematics to concern *possibilia* is also supposed to render more plausible the logical status granted to mathematical truths by Tarski's account.[16] Etchemendy's response here is sufficient: "I'm not sure which is more difficult to accept, the idea that we can build structures out of non-existent objects or the idea that Tarski had this in mind" [Etchemendy, 2008, 286].

In any case, the move is otiose as long as within the set theory in which we express the model theory every model constructed from the urelements of any possible world is isomorphic to some model constructed from pure sets [McGee, 1992, 275–8], since from this it follows that for any structure that might be a model some set of sentences there is a structure that actually is. The comfort of thinking that on Tarski's account logic only tells us how many things might exist, rather than how many things actually do, is therefore lost.

If one doesn't go in for attributing a modal aspect to Tarski's consequence relation on these grounds, the other possible strategy is to downplay the force of Tarski's "must". In fact, of course, the two strategies blend into one another; witness Sher's equation of "formal possibility" with "mathematical existence" at the passage cited above.

An example of attempting to downplay the force of Tarski's "must" is Ray's reading:

> I think that in every case the modal terms in question can be construed as a familiar colloquial use of modal locutions, or as indicating nothing stronger than deductive consequence ... I think that Tarski is simply using the logician's "must" indicating the logical necessity of the consequence, not the consequent [Ray, 1996, 654].

This begs the question against Etchemendy: the whole issue is whether Tarski's definition supports any notion of "logical necessity". If deductive consequence has modal features, either alethic or epistemic, and if there is some question as to whether Tarski's definition captures these features, as there is, then one doesn't defend Tarski by simply taking it as given that the relation he has defined is the relation of "deductive consequence". Ray's nervous "logical necessity, if you will" [Ray, 1996, 654] indicates that not even Ray himself is particularly convinced by this slide. As Etchemendy shows quite well, whatever it is Tarski defines, it lacks basic features that informal glosses on consequence like "if the premises are true, the conclusion must be true" attribute to the consequence relation.[17]

Nevertheless the question remains of what Tarski thinks can be proved, and also of what he thinks "must" means in both condition (F) and the argument we have been considering that condition (F) holds of a consequence pair if Tarski's definition does.

7.4.2   Consequence and truth

The "fallacy" reading of the passage and the responses we have surveyed assume that what Tarski claims can be proved is that the consequences of true sentences *must be* true. Here is my suggestion: Tarski isn't claiming that at all; the point is rather that it can be proved that the consequences of *true* sentences must be *true*. The proffered argument, then, isn't intended to ward off the suggestion that Tarski's account implies merely that the consequences are true and not, further, that they "must be" true; it is intended to establish that the intuitive connection between consequence and truth is captured by the definition. The contrast here is Carnap's conception of L-consequence, which is characterized in terms of "analyticity" in the sense of following from everything by Carnap's transformation rules. As we have already seen, Tarski wants to replace analyticity in Carnap's approach by truth, since he thinks that Carnap's objections to semantic notions are bad and that he himself has formulated the means to make attributions of truth acceptable by Vienna Circle standards.

In support of this, we can note that condition (F) requires exactly that consequences of a set of sentences be true if the members of the set are. Condition (F), in turn, is motivated by the consideration that "from the point of everyday intuitions, it is clear that it cannot happen that all the sentences of the class $\mathfrak{K}$ would be true but at the same time the sentence X would be false" [Tarski, 2002, 183]. Moreover, the validity of the ω-rule is characterized the same way: the failure of derivability conceptions of consequence is that they don't declare all ω-inferences valid, yet of these inferences, with their premises A_0, A_1 ... and universally quantified conclusion A, Tarski says that, according to the "everyday concept", "whenever all of these sentences are true, the sentence A must be true". According to Tarski, then, the "everyday concept" that is to be captured by his definition is characterized in terms of truth. Thus what Tarski claims for his definition is not that it captures the modal features of the "common" concept of consequence (more on this later), but that it corrects Carnap's conception of L-consequence by reinstating the ordinary connection between consequence and truth. This follows easily, so the "proof" can be omitted, but it bears emphasis in the context of a response to Carnap.[18]

That this aspect of his defined consequence relation is of interest to Tarski addresses the question sometimes raised (e.g. [Bays, 2001, 1718]) as to why Tarski takes it that on a good account of consequence the conclusions of ω-inferences are consequences of their premises and that Gödel sentences for theories are consequences of these theories. We'll come back to the ω-rule, but in the latter case the connection is clear: the whole point of a Gödel sentence is that it is true if the theory for which it is a Gödel sentence is true, so, if a sentence is a consequence of others if, if they are true, it is true, then Gödel sentences are consequences of the theories for which they are Gödel sentences. This isn't, of course, the standard take on Gödel's results, but both Carnap and Tarski take, in the immediate aftermath of Gödel's results, the interesting line that what Gödel has shown is not that there are sentences and theories such that the sentences are true if the theories are true even though these sentences aren't consequences of the theories, but that consequence isn't an effective notion—this is why Carnap feels forced in light of Gödel's results to introduce "indefinite" transformation rules. Tarski, working essentially from Carnap's approach, accepts this (cf. [Etchemendy, 1988, 72], but wants to reintroduce the intuitive link between consequence and truth.

Thus Tarski was not guilty of "Tarski's fallacy"—not, however, because he gave an argument that in an inference declared logically valid by his definition, necessarily if the premises are true so is the conclusion that was in fact valid, but because he offered no argument for any conclusion of the sort. He did, of course, use common locutions like "must be true" and we will have to look into what he could have meant by this.

7.4.3 Tarski's "must"

Below we will consider the intuitive adequacy of Tarski's account. Here the question is what Tarski himself meant by the "must"s in condition (F) and in his comments on it and on his definition. The key here, again, is the fact that Tarski is attempting simply to modify Carnap's definition of L-consequence. As Tarski writes, what he wants is a treatment on which logical consequence is "a relation which is to be uniquely determined by the form of the sentences between which it holds" and on which:

> this relation cannot be influenced in any way by empirical knowledge, and in particular by knowledge of the objects to which the sentence X or the sentences of the class K refer [Tarski, 1983h, 414–5].[19]

Whatever one's opinion of this as an attempt to capture the intuitive meaning of "follows logically", there is something modal about the characterization: logical truths are truths independent of the semantics of non-logical terms. "Must be" then means that logical consequence is independent of empirical knowledge, in the sense that it is independent of the semantics of the language to the extent that this involves objects that are only known empirically. The notion is at least "modal" in the sense that whether or not a consequence-pair $(\mathfrak{K}, X)$ instantiates the consequence relation isn't settled by the extensional semantics of the sentences in $\mathfrak{K}$ and X alone, but depends also on the semantic values of other expressions throughout the language. Notice in particular that Tarski's construction here matches Carnap's treatment of analyticity and L-consequence: logical truths are supposed to be analytic, following from the empty set, and hence not depending on "empirical" knowledge [Carnap, 2002, 180], and they are supposed to be independent of the meanings of non-logical terms, in accord with Carnap's treatment of L-consequence [Carnap, 2002, 180–181]. It is only to be conceded that this "modality" has nothing to do with supporting counterfactuals and is not epistemic in that inferences might be logically valid or invalid without our being able to know that they are since validity depends on the distribution of truth values across sentences of the language not involved in the argument. On Tarski's Carnapian conception "if the premises are true, the conclusion must be true" means "either the conclusion is true or a premise is false, no matter what the interpretation of the non-logical vocabulary".

We will return to Tarski's conception of form below; the point here is that when Tarski says that the consequences of true sentences must be true, he simply follows Carnap in meaning that they are true no matter what are the semantic values of non-logical expressions in the language: $(\mathfrak{K}, X)$ is valid if it is materially valid on every uniform interpretation of the non-$\mathfrak{F}$ vocabulary it involves. Tarski does have a substantial, intensional conception behind the "must" in condition (F); it is just that we must concede that it doesn't capture any interesting epistemic or alethic modality.

7.5 The formality problem and the logical constants

So far we have seen that Tarski's account cannot capture the modal features intuitively involved in logical consequence, whether these are epistemic or alethic, and that Tarski's account ranks many mathematical truths as logical truths.[20] Common to both problems is an apparent mismatch between intuitive *definiens* and Tarskian *definiendum*: Tarski's definition both fails to capture the ordinary term's modal features and seems to get its extension wrong. In the context of concern about these problems, it is tempting to interpreters to appeal to the other aspect of logical consequence Tarski mentions just before setting out his definition, its "formality".

7.5.1 Constant and consequence

The issue here comes down to Tarski's appeal to the idea that truths and inferences are formal when they involve a certain distinguished vocabulary. As Tarski himself notes, "Underlying our whole construction is the division of all terms of the language into logical and extra-logical" [Tarski, 1983h, 188]. This raises the hope that the otherwise apparent failings of the definition could be corrected by the proper selection of logical vocabulary. Etchemendy (as in other places, substantially anticipated by Kneale: [Kneale, 1961, 96–7]) writes:

> Since Tarski's general account captures none of the modal, epistemic or semantic characteristics of logical truth and logical consequence, it seems that these characteristics must somehow emerge from the sought-after supplement, from our account of what makes certain expressions "genuinely logical" and others not.

This is why the task of characterizing the logical constants comes to seem at once so important and yet so difficult. Indeed, most of the burden of Tarski's analysis seems to shift to exactly this issue. But by now it should be clear that the issue is based on a confusion—namely, the assumption that when the account works, it works due to some peculiar property of the expressions included in $\mathfrak{F}$. But this assumption is false: there is no property of expressions that guarantees the right extension in these cases, none whatsoever. After all, any property that distinguishes, say, the truth functional connectives from names and predicates would still distinguish these expressions if the universe were finite. But in that eventuality, Tarski's account would be extensionally incorrect. This observation alone is enough to show that it is not any property of the expressions we hold fixed, the so-called logical constants, that accounts for the occasional success of Tarski's definitions [Etchemendy, 1990, 128].

Since, as Tarski notes, the whole account depends upon a principled identification of the logical terms, the task of interpreters who want to claim that Tarski's analysis is substantially right has been to claim that there is a distinguished set of logical expressions, that these expressions are indeed intuitively "formal", and that when $\mathfrak{F}$ is taken to be this set of expressions, the account at least doesn't overgenerate and even perhaps captures the modality intuitively involved in consequence after all. Such defenses generally also require a defense of Tarski's identification of at least some mathematical truths as logical truths, especially given that "=" is generally taken to be logical on current views of logic.

Our way can be short with the suggestion that a principled identification of the constants would help with the modality problem. It is manifest that on *any* selection for $\mathfrak{F}$ the account captures no appropriate alethic modality, since it generalizes only over actual semantic values. It is equally manifest that on no selection for $\mathfrak{F}$ does it capture any appropriate epistemic modality, since knowing whether an inference is valid would require, by the definition, knowing the distribution of truth-values across a host of other sentences in the language. Whether, in turn, the move helps with overgeneration depends not so much on the details of the conception of the logical constants offered, but on the concomitant defense of the logicism involved in that selection.[21] That is, though of course the plausibility of the logicism involved depends on the selection of the constants, the real weight of the account rests on the defense of the claim that *any* mathematical truths are logical, and not on the details of which ones come out logical for the mooted selection of logical constants. Since this is so we won't tarry long here over the question as to whether a principled account of the constants would help with the modality and overgeneration problems, and we will confine our attention simply to attempts to supply some criterion of logicality on Tarski's behalf.

Interpreters who want to claim that Tarski actually endorses the suggested approach must also read his skeptical remarks at the end of the article as merely the announcement of a research program:

> At the foundation of our whole construction lies the division of all terms of a language into logical and extra-logical. This division is certainly not entirely arbitrary: if we did not count among the logical terms e.g. the implication sign or the quantifiers the definition provided of following could lead to consequences manifestly contrary to everyday intuitions. On the other hand however I know no objective reasons which would allow one to draw a precise dividing line between the two categories of terms ...
>
> Clearly, further investigations may throw a lot of light on the question which interests us; perhaps one will succeed with the help of some weighty arguments of an objective character in justifying the dividing line traced by tradition between logical and extra-logical terms. Personally I would not be surprised however even if the result of these investigations were to be decidedly negative and if hence it would turn out to be necessary to treat such concepts as following logically, analytic sentence or tautology as relative concepts which must be related to a definite but more or less arbitrary division of terms of a language into logical and extra-logical; the arbitrariness of this division would be in some measure a natural reflection of that instability which can be observed in the usage of the concept of following in everyday speech [Tarski, 2002, 188–90].

I think it is fairly clear from the wording that this expresses skepticism, and I will substantiate this in what follows, but many interpreters take the passage the other way. The primary way to supply Tarski with an account of the constants is to read the view Tarski's much later "What are Logical Notions?" of 1966 back into the 1935 article, and it is to readings based on this maneuver that we now turn.

7.5.2 Anachronistic readings

The conception of logic in the 1966 lecture is now familiar. Tarski refers to Klein's *Erlanger Programm* for the classification of geometries in terms of transformations, and notes that that the concepts of e.g. affine geometry and topology may be distinguished by the transformations under which their extensions are invariant. The more transformations of the universe of discourse onto itself one allows, the fewer are the properties of objects that are invariant under the transformations, and hence the fewer and more general are the "notions" involved. Tarski then enters his suggestion:

> Now suppose we continue this idea, and consider still wider classes of transformations. In the extreme case, we would consider the class of *all* one-one transformations of the space, or universe of discourse, or 'world' onto itself. What will be the science which deals with the notions invariant under the widest class of transformations? Here we will have very few notions, all of a very general character. I suggest that they are the logical notions, that we call a notion 'logical' if it is invariant under all possible one-one transformations of the world onto itself [Tarski, 1986j, 149].

Since here Tarski simply identifies "notions" with extensions [Tarski, 1986j, 147] we are already far from the concerns of the 1930s. Tarski closes the article by noting that everything definable in *Principia Mathematica* is logical in the defined sense, but that the question of whether mathematics is a branch of logic can only be decided if it is decided whether such a type-theoretic conception of logic, as opposed to first-order logic supplemented by a set theory in which the membership relation is non-logical, is correct, a question Tarski declines to answer.

There is a significant literature on Tarski's invariance criterion, e.g. [Sher, 1996], [Sher, 2008], [Feferman, 1999], [Feferman, 2010], [McGee, 1996], [Gómez-Torrente, 2002], [Bellotti, 2003], [Casanovas, 2007], [Hanson, 1997]. We needn't enter into it, however, because the definition of a constant offered in the 1966 lecture simply has nothing to do with the notion of formality at work in [Tarski, 2002]. Admittedly, the invariance criterion of 1966 is related to Theorem I of [Tarski, 1983i], but the very fact that in 1935:

> Every relation between objects (individuals, classes, relations, etc.) which can be expressed by purely logical means is invariant with respect to every one–one mapping of the 'world' (i.e. the class of all individuals) onto itself [Tarski, 1983i, 385].

is a *theorem* rather than a definition of logicality gives the game away.

Rather, in [Tarski, 2002] Tarski quite forthrightly retains Carnap's conception of a logic as formal in the sense of being concerned only with the forms *of sentences*. The 1966 article, by contrast, understands logic in terms of the "form" of the universe of discourse—in terms, that is, of mathematical structure. Whatever the value of such an understanding of logic, it can't have anything to do with what Tarski is interested in in [Tarski, 2002], since in the earlier article, as we have now seen in some detail, Tarski's aim is merely to supplement Carnap's approach with his semantics.

Since this is so we needn't enter further into the details of such readings. Foremost here are Sher's works [Sher, 1991], [Sher, 1996], [Sher, 2008], and we can note that Sher doesn't really suggest that supplementing [Tarski, 2002] with [Tarski, 1986j] produces historical accuracy. If our goal is to understand what Tarski thought about logical constancy in 1935, looking forward to 1966 is not the answer.

7.5.3   Carnap on formality

Unlike the anachronistic readings of Tarski, a reading of the article in terms of *Logical Syntax* is without problems. Return again to Tarski's important remarks on the consequence relation at [Tarski, 2002, 183–4] cited above. Four properties are attributed to the consequence relation by this passage:

1. it holds between sentences in a way uniquely determined by their form
2. it is independent of empirical knowledge
3. it is independent of the semantic values of expressions in the type-theoretic hierarchy
4. it is invariant under substitutions of vocabulary within the type-theoretic hierarchy

All of these features are straight out of *Logical Syntax*. The fourth is simply Carnap's criterion of L-consequence as stated at [Carnap, 2002, 181], and the third is Tarski's generalization of the same idea. The first is also consonant with Carnap's "syntactic" approach, and the second makes the contrast between logic and the empirical that Carnap stresses repeatedly:

> If a material interpretation is given for a language S, then the symbols, expressions, and sentences of S may be divided into logical and descriptive, i.e. those which have a purely logical, or mathematical, meaning and those which designate something extra-logical—such as empirical objects, properties, and so forth. This classification is not only inexact but also non-formal, and thus is not applicable in syntax. But if we reflect that all the connections between logico-mathematical terms are independent of extra-linguistic factors, such as, for instance,

empirical observations, and that they must be solely and completely determined by the transformation rules of the language, we find the formally expressible distinguishing peculiarity of logical symbols and expressions to consist in the fact that each sentence constructed solely from them is determinate. [Carnap, 2002, 177].

Tarski's preferatory remarks to condition (F) simply recount the last part of this passage.[22] Tarski's conception of logicality in [Tarski, 2002] is simply Carnap's. Not surprisingly, then, it shares its weaknesses; one question here was what, exactly, Tarski's attitude was toward these weaknesses.

7.5.4   The ω-rule and Gödel sentences

A central topic in the reading of [Tarski, 2002] which is of relevance to both the overgeneration and formality problems is the treatment of the ω-rule at [Tarski, 2002, 177–81]. Having noted the weaknesses of conceptions of consequence in terms of finitary proof rules at the outset of [Tarski, 2002], Tarski returns to the topic of the ω-rule from § 5 of CTFL and [Tarski, 1983l]. Tarski notes that he has given examples of ω-incomplete theories that are "quite elementary". Of the existence of such theories, he writes:

> This fact attests by itself, I think, that the formalized concept of following, which until now was generally used in the construction of deductive theories, by no means coincides with the everyday concept—after all, from the point of view of everyday intuitions it seems indubitable that the sentence A follows from the totality of sentences $A_0 \ldots A_1 \ldots A_n \ldots$: whenever all these sentences are true, then also the sentence A must be true [Tarski, 2002, 178].

Etchemendy, assuming that—as [Tarski, 2002, 178] read most flatly indicates—Tarski is thinking of the ω-inference he discusses as formulated in a first-order theory with numerical constants, complains that the example misfires: first-order arithmetic is ω-complete neither proof-theoretically (that's what Gödel's incompleteness theorem shows) nor model-theoretically (that's what the incompleteness theorem plus the completeness theorem for first-order logic shows) [Etchemendy, 1988, 73]. As a result, Etchemendy argues, the only way that Tarski can get the ω-rule to come out valid is to stipulate that the numerals are logical constants, but, the suggestion is, this is an *ad hoc* selection of constants.

The standard response in the literature here is right as far as it goes, but suffers as much as the rest of the literature from ignoring Tarski's relationship to *Logical Syntax*. As many commentators have pointed out (e.g. [Gómez-Torrente, 1996], [Bays, 2001]) Tarski's other discussions of the ω-rule assume that the inferences in question are formulated in STT with numerical terms defined in the logicist manner. As these commentators note, the ω-rule is valid so formulated in STT and the selection of constants involved isn't *ad hoc*. (There are disputes of detail over the rendering of the ω-rule the reading assumes, e.g. [Edwards, 2003], [Sagüillo, 1997], but we needn't enter into these here.)

This, again, is right as far as it goes but it neglects the fact that Carnap formulates the ω-rule in Language I with the successor-stroke and the numerals stipulated to be logical [Carnap, 2002, 16,25]. Since Tarski has Carnap in mind elsewhere in the article, his formulation of the ω-rule with numerals could well simply be as Etchemendy has it (*contra* [Bays, 2001, 1721])—with the difference that treating the numerals as logical is not *ad hoc* but rather simply derived from Carnap's presentation.[23]

A more general worry concerns Tarski's equally enthusiastic treatment of Gödel sentences as consequences, in his sense, of the theories for which they are Gödel sentences. As Etchemendy suggests [Etchemendy, 1988, 73], this requires taking nearly all of the vocabulary of the relevant theories as logical. Though the point stands that Gödel's results apply to any theory that extends a weak system of arithmetic and hence that, in principle, a Gödel sentence could involve just about any vocabulary one pleased, if we take Tarski's attention to have been on Gödel sentences for plausibly mathematical theories (as the discussion of proof-theoretic encodings of the ω-rule that precedes the discussion of Gödel sentences at [Tarski, 2002, 180–1] indicates) then the response is that Tarski already counted this vocabulary as logical anyway, either by following Carnap's treatment of Language I or by taking the theories to be formulated in STT. For more far-flung theories (e.g. an extension of arithmetic that includes "is a cow") Gödel sentences *could* involve intuitively non-logical vocabulary, but this vocabulary will have found its way into the Gödel sentence only by being caught up in the construction of the provability predicate for the theory and, plausibly, its having done so will reflect a decision on the part of the theory-builder to treat the vocabulary as logical. Thus there isn't really any additional problem here for Tarski.

7.5.5   Antitractarianism and the nature of logic

All that said, Tarski's treatment, no less than Carnap's, founders on the assumed role of the logical constants—*if* we take [Tarski, 2002] as seriously intended to offer a conception of logical consequence that is supposed to be supplemented by some definitive account of the logical constants. On my reading, that no such account is available is actually part of the point. As I have suggested earlier, Tarski's main aim in the article, beyond supporting the semantics advertised in ESS by improving upon Carnap's work is to undermine the Tractarian conception of logic to which Carnap was still attracted. For this purpose, simply raising the need for a criterion of logicality and emphasizing the unlikeliness that one would be found was sufficient. Since Tarski's view was the Warsaw view on which there weren't principled distinctions between logic and mathematics, or between those and science [Murawski, 2004], [Woleński, 1989], [Carnap, 1963, 30], Tarski in the lecture isn't hoping for a criterion of logicality, but is rather pointing out that none will be found.

As we noted above, Carnap in *Logical Syntax* is thoroughly convinced that the doctrine that logic "says nothing about facts" plus a treatment of mathematics as "logical" is the only way to a satisfactory account of the role of mathematics in science. If logical and mathematical claims "say nothing about the facts" then traditional epistemological and metaphysical questions about mathematical facts and objects and how we know them can be set aside. The whole construction, however, depends on a principled delimitation of logic and an argument that logic so delimited has the needed features.

From their earliest meeting, Tarski had disputed these views. Here are Carnap's notes on their meeting of February 22, 1930:

> 8–11am with Tarski in the café. Of monomorphism, of tautologies, he will not admit that that they say nothing about the world: he says that there is only a gradual and subjective distinction between tautological and empirical sentences (quoted by Haller at [Tarski, 1992, 5]; my translation).

Carnap recounts this disagreement in his "Intellectual Autobiography" as well [Carnap, 1963, 30]. In Tarski's letter to Neurath of September 7, 1936, we have this recollection:

> I recall a conversation with Carnap which I had a year ago in Prague (during which also Mrs Lutmann and, it seems to me, Mrs Carnap were present). Namely, at my request Carnap told wherein consisted the influence of Wittgenstein on the Vienna Circle and on him personally. He said then, that this influence was on the one hand challenging, on the other, hindering. The influence was challenging, for W. focused attention on the importance of problems which concern the language, for he brought to light the reduction of "philosophical" problems to linguistic problems; the influence was however also hindering, for just

W. contested and rejected the possibility of speaking about a language in the legitimate and fault-free manner. And now Carnap was of the opinion, that just the liberation from this hindering influence of W. is due to the Varsovians (and especially to my Vienna lectures) [Tarski, 1992, 25].

So during the Prague *Vorkonferenz* to the Paris Unity of Science congress Tarski had been interested to know what Wittgenstein's influence was, and had been told by Carnap that the doctrine that one couldn't speak about language was a retarding influence of Wittgenstein. Tarski was surely pleased to be credited by Carnap with having convinced him that serious theory about language is scientifically acceptable. Notice here, however, that what Carnap *doesn't* say in 1934 is that *talk about relations between language and the world* is acceptable. That is, in 1934 Carnap credits Tarski only with as much as is embodied in *Logical Syntax*, and attributes this to Tarski's lectures of 1930, which concerned Tarski's work on consequence—work which is about language but not about language–world relations. Having developed semantics in the meantime, Tarski would have wanted to draw Carnap along with him that much further, and it is to this end that ESS and CFL are pitched. Tarski, then, wants to undermine the distinction betwen logic and science that continues to be enshrined in Carnap's definition of L-consequence, and the distinction between L-consequence and consequence in general.[24]

Tarski's own wording at the end of [Tarski, 2002] hardly suggests endorsement of the possibility of a principled delimitation of the logical constants, since Tarski says that he "would hardly be surprised" [Tarski, 1992, 189] if no such distinction could be made in a principled way and, moreover, Tarski explicitly characterizes the purport of the passage as skeptical in a contemporaneous letter to Neurath:

> I gave a lecture in Paris about the concept of logical consequence; there I contested (among other things) the absolute character of the division of concepts into logical and descriptive as well as of sentences into analytic and synthetic, and I endeavored to show, that the devision of concepts is quite arbitrary, and that the division of sentences has to be related to the division of concepts [Tarski, 1992, 28].

Here the distinction between analytic and synthetic sentences and the delimitation of the logical constants are "quite arbitrary" and Tarski "contests" the absolute character of the putative distinctions; matters could hardly be clearer.

In denying any principled distinction between logic and mathematics, or between them and science, Tarski was simply following the Lvov–Warsaw school. Murawski writes that "The semantic point of view implied the rejection of the analytical concept of logic, i.e., the rejection of the thesis that logic is a collection of tautologies that are contentually empty (this is a thesis on logic and mathematics versus reality). Leśniewski (and Kotarbiński) claimed that logic describes the most general features of being, logic plays a role of a general theory of the real world" [Murawski, 2004, 330–1], but unfortunately provides no citations. Looking to the texts, Kotarbiński, though he spills some ink over the doctrine that logic concerns the "form" of reasoning is happy to call Leśniewski's "Ontology" a "general theory of objects" [Kotarbiński, 1966, 211].[25] Looking over the main texts of Kotarbiński, Leśniewski, Łukasiewicz and Ajdukiewicz, one does notice a conspicuous absence of worry about the status of mathematics—the worry that otherwise dominated analytical philosophy from Frege onwards.

Granting that, the Lvov–Warsaw school was in direct conflict with the Vienna Circle on this point; see statements such as this one from Hans Hahn:

> If logic were to be conceived—as it actually has been conceived— as a theory of the most general properties of objects, as a theory of objects as such, then empiricism would in fact be confronted with an insuperable difficulty. But in reality logic does not say anything at all about objects; logic is not something to be found in the world; rather logic first comes into being when—using a symbolism—people *talk about the world*, and in particular, when they use a symbolism whose signs do *not* (as might at first be supposed) stand in an isomorphic one–one relation to what is signified [Hahn, 1980, 40].

Though there is a small industry devoted to supporting Tarski's supposed commitment to an account of the distinction between logical and non-logical expressions, matters could hardly be clearer: Tarski never believed in any such distinction. The evidence from the letters to Neurath [Tarski, 1992] is compelling here, and the concluding remarks to [Tarski, 1986j] show that he never changed his mind on the topic.

What *is* worthy of note is that Tarski brings this skepticism to bear in the formulation of a conception of consequence much more at home in the Vienna Circle conception of logic than that of the Lvov–Warsaw school. Carnap's conception of L-consequence sits resolutely within the framework of the Tractarian conception of logic. Tarski takes it back out of that conception when substitution is replaced with semantics and provability with truth, but in this context the view that there is no radical discontinuity between logic and science takes the form of skepticism about the possibility of a conception of the constants that would support the Vienna Circle conception of logic.

In this regard it is not surprising that when Tarski 30 years later does formulate a criterion of logicality in a constant, it is an "ontological" one, accompanied again by emphasis on the continuities running from logic through mathematics to science; see here especially the discussion of "∈" in the conclusion of [Tarski, 1986j]. The later article makes good on the suggestion of the earlier one that no non-representational understanding of logicality will be found by proposing a characterization of logic that is unabashedly representational: logical terms are simply those that denote general structural features of the universe of discourse. We can note also, in the context of Tarski's response to the Tractarian doctrines to which Carnap was attracted, that the 1966 lecture upholds exactly the conception of logicality as generality that Wittgenstein criticized in Russell (e.g. *Tractatus Logico-Philosophicus* 6.1231, 6.1232).

7.6 Evaluating Tarski's account

7.6.1 The analytic problem

With all of the foregoing in hand, we can ask ourselves a basic question about Tarski's conception of what he was doing. Since Etchemendy is quite right that logical consequence as defined by Tarski lacks many features of the ordinary notion of "validity" or "following from", we need to ask ourselves what was the relation Tarski conceived to hold between his analysis and the ordinary notion. Now, Tarski begins the article by insisting that the connection is important:

> The concept of *logical consequence* is one of those whose introduction into the field of strict formal investigtion was not a matter of arbitrary decision on the part of this or that investigator; in defining the concept, efforts were made to adhere to the common usage of the language of everyday life [Tarski, 1983h, 409].

And just after offering his definition, he writes:

> It seems to me that everyone who understands the content of the above definition must admit that it agrees quite well with common usage [Tarski, 1983h, 417].

The phrasing here corresponds quite well to the remarks in other passages where Tarski introduces a definitional task—for instance the remarks on "definable sets of real numbers" in [Tarski, 1983f] and of course on truth in [Tarski, 1983a]. In view of the work we have already done on intuitive meaning and deductive theories in Tarski the strong default assumption is that Tarski was doing in [Tarski, 2002] exactly what he was doing in these other works, showing how to express an "intuitive", "everyday" concept within a deductive theory.

Nevertheless the gross divergence between what Tarski offers and the features that Etchemendy rightly stresses has tempted some commentators to propose that Tarski simply wasn't trying to capture the "common usage of the language of everyday life". Jané, for instance, writes:

> Tarski's common concept is not some general, all-purpose notion of consequence, but a rather precise one, namely the concept of consequence at play in axiomatics [Jané, 2006, 2–3]

where "axiomatics" was the study of axiomatic systems following Hilbert and especially the Peanists, Pieri in particular. The attraction here is obvious: since Tarski clearly doesn't capture essential features of the ordinary notion, one proposes that he wasn't trying to capture the ordinary notion.[26] However, Tarski is trying to capture the ordinary notion—there are just too many passages in the article that say this, and capturing ordinary notions with expressions constrained by the structure of deductive theories is what he had been trying to do all along.

And what should our own take on Tarski's conception of consequence be? The modal and extensional failings of the analysis are manifest. No account of the logical constants will correct this, though acceptance of something like logicism will at least blunt the force of overgeneration.

Why does Tarski accept Carnap's conception, despite its obvious flaws, and despite its poor fit with the Lvov–Warsaw school's generality conception of logic? First, the extensionalism of the Lvov–Warsaw school would prevent the inclusion of any genuinely modal notion in the definition of consequence anyway. In this respect Tarski simply couldn't have been bothered by what to Etchemendy, and to most of us, are the clear failures of a conception of consequence that is, at bottom, material. But why, we can still ask, was Tarski attracted to Carnap's conception of consequence? Here the history matters: Carnap's conception simply *was* an improvement over the going derivational conceptions of consequence that Tarski had accepted until sometime a couple of years after Gödel's results and which we can find in Kotarbiński, Leśniewski and others among his teachers. As we have seen, Carnap reacted to Gödel's results not by holding that they showed that there were truths of the language of a theory which weren't consequences of it, but that there were consequences of many theories that aren't derivable (cf. [Gómez-Torrente, 1996, 146]). Though not the contemporary gloss on the results, given that Gödel sentences are true if the theories for which they are Gödel sentences are, and given a basic "everyday" gloss on consequence that Tarski's semantics had put him in a position to accept, namely that a consequence of a set of sentences is a sentence that is true if they are, this was a reaction to Gödel that would have seemed natural to Tarski, and that would have led, precisely by Tarski's taking himself to have tamed semantic notions for serious theory, to his conception of consequence.

7.6.2 Eliminating transformation rules

Carnap's approach to logical vocabulary, as is often noted, comes in two forms. In the treatments of Languages I and II the logical expressions are simply listed. In "general" syntax, Carnap's criterion of logicality is as above: logical expressions are those such that every sentence involving only them is either provable or refutable on no assumptions, given the transformation rules of the language. Now the the most important aspect of this is that it retains an essential reference to the transformation rules. General syntax is general in that it makes claims about the relationships among transformation rules and other aspects of a language, but all of its notions are nevertheless thereby relative to a choice of transformation rules, and about these Carnap has nothing to tell us; indeed the whole point of the Principle of Tolerance is that nothing of a critical nature can be said about the choice of transformation rules.

It is this focus on stated (if, in General Syntax, generalized over) transformation rules syntactically conceived that weakens Carnap's response to Gödel in two respects, both of which are corrected by Tarski's semantic treatment. First, since Carnap demands that everything be done in terms of syntax, he cannot state a common intuitive upshot of the first incompleteness theorem, namely, that there are sentences that are true in the language of any extension of a weak arithmetic that don't follow from the axioms. The only way he has to say that anything that comes close to the forbidden talk of truth to this is to say that something is a "consequence" of the axioms. So the transformation rules have to be strong enough to state Gödel's result as the claim that there are "consequences" of the axioms that nevertheless aren't derivable, and the ω-rule is thereby required.

Second, and more generally, if the aim of the enterprise is to capture the ordinary notion of one thing "following from" another then Tarski's treatment, whatever other weaknesses it shares with Carnap's, is one respect a major improvement precisely because it is no longer relativized to a set of transformation rules, a relativization simply not to be found in ordinary thought about consequence. Ultimately Carnap is still stuck in the predicament of being unable to do anything more in the treatment of logical consequence than to assume a list of formal rules. With the move to tolerance these rules are no longer supposed to be intuitively valid, but Carnap fails to replace that appeal to intuitive validity with any other analytic connection to ordinary thought about consequence, beyond various officially forbidden appeals to ordinary thought about truth and consequence. Since Languages I and II do happen to have transformation rules that are, if complicated in places, apparently valid once stated, Carnap's various definitions of "analyticity" and his pumping of the intuition that the criterion of "L-consequence" matches intuitions about what is true "in a purely logical way" at [Carnap, 2002, 180–1] give the reader a misleading sense of security that "analytic" sentences are really in some sense analytic, and "L-consequences" are really in some sense logical consequences. Imagining Carnap's definitions applied to a language with consistent but intuitively bizarre transformation rules reminds us that "analyticity" and "L-consequence" only coincide with analyticity and logical consequence in fortuitous cases. In the context of a debate with intuitionists and finitists the Principle of Tolerance seems acceptable since all of the competing systems have at least some intuitive backing. We should not, however, mistake tolerance between choices all of which have some intuitive backing for a proof that intuitive backing isn't required.

We can see some of the effects of Tarski's elimination of the transformation rules in his remark that logical consequence degenerates to material consequence when all sentences of the expressions of the language are included in $\mathfrak{F}$. Carnap's conception of L-consequence does not degenerate to material consequence if all expressions of the language are logical. To see this, consider a language with a single transformation rule according to which everything follows from everything. All sentences of this language are "analytic" and all its expressions are "logical", but it surely isn't a language in which consequences of true sentences are true.

As we have seen, Carnap first introduces the notion of consequence in general in terms of transformation rules; he then distinguishes L-consequence as consequence that is preserved under uniform substitutions of descriptive vocabulary [Carnap, 2002, 180–1]. The difference from Tarski is that a "transformation rule" according to which one sentence follows from others just in case either it is true or one of them is false could not have been contemplated by Carnap, since it goes beyond the bounds of "syntax" in being formulated in terms of truth rather than the structural features of sentences. (Here "analytic" can't be substituted for "true", since "analytic" is itself defined by what follows according to the transformation rules.) So in mentioning the degeneration of logical into material consequence when all expressions are "logical", Tarski is not commenting on Carnap's view, but pointing out a consequence of his modification of it. The dependence on transformation rules in Carnap's view helps artificially to support the Vienna Circle's Tractarian distinction between sentences that "say nothing about reality" and sentences that do; in the material consequence passage Tarski shows that his conception of consequence undermines the distinction and reinstates the continuity of logic, mathematics and science that was accepted in the Lvov–Warsaw school.

Tarski elsewhere characterizes semantics in terms of "the totality of considerations concerning those concepts which, roughly speaking, express certain connexions between the expressions of a language and the objects and states of affairs referred to by these expressions" [Tarski, 1983b, 401]. The article is an advertisement for Tarski's methods. Notice that, in the terms of ESS, one way of resisting the importance of "scientific semantics" would be to rest on a Tractarian distinction between sentences that do and those that do not bear "connexions" to objects and states of affairs. Read as a pair, ESS and CLC make an extended argument in favor of Tarskian semantics: the first article sketches the basics of how it works, while the second undoes the primary source of resistance he expected from his audience, namely, Tractarian doctrines about logic as they were found in the Vienna Circle. In particular, Tarski's account of consequence in [Tarski, 2002] makes logical consequence a semantic concept in the sense of [Tarski, 1983b, 401], since consequence is defined in terms of models, which are defined in terms of the explicitly semantic concept of satisfaction.

7.6.3   Epistemic and generality conceptions of logic

Intuitionistic Formalism centrally involved an epistemic conception of logical consequence. Recall the passage from Leśniewski that Tarski cites at [Tarski, 1983c, 61]:

Having endeavored to express some of my thoughts on various particular topics by representing them as a series of propositions meaningful (*sinnvoller Sätze*) in various deductive theories, and to derive one proposition from others in a way that would harmonize with the way I finally considered intuitively binding (*welche Ich "intuitiv" als für mich bindend betrachte*), I know no method more effective for acquainting the reader with my logical intuitions ("*logischen Intuitionen*") than the method of formalizing any deductive theory to be set forth [Leśniewski, 1992e, 487] [Leśniewski, 1929, 78].

In this respect Intuitionistic Formalism involved an account of consequence that Aristotle himself, in his account of the perfect syllogism at 24b18ff, could have recognized:

> A syllogism is discourse in which, certain things being stated, something other than what is stated follows of necessity from their being so ... I call that a perfect syllogism which needs nothing other than what has been stated to make plain what necessarily follows.

As Etchemendy notes, Tarski's account simply lacks the relevant features:

> Surrounding the intuitive concepts of logical consequence and logical truth are a host of vague and philosophically difficult notions—notions like necessity, certitude, a prioricity, and so forth. Among the characteristics claimed for logically valid arguments are the following: If an argument is logically valid, then the truth of its conclusion follows necessarily from the truth of the premises. From our knowledge of the premises we can establish, without further investigation, that the conclusion is true as well. The information expressed by the premises justifies the claim made by the conclusion. And so forth. These may be vague and ill-understood features of valid inference, but they are the characteristics that give logic its *raison d'être*. They are why logicians have studied the consequence relation for over two thousand years ...

> The property of being logically valid cannot simply consist in membership in a class of truth preserving arguments, however that class may be specified. For if membership in such a class were all there were to logical consequence, valid arguments would have none of the characteristics described above. They would, for example, be epistemically impotent when it comes to justifying a conclusion. Any uncertainty about the conclusion of an argument whose premises we know to be true would translate directly into uncertainty about whether the argument is valid. All we could ever conclude upon encountering an argument with true premises would be that either the conclusion is true or the argument is invalid. For if its conclusion turned out to be

false, the associated class would have a non-truth-preserving instance, and so the argument would not be logically valid. Logical validity cannot guarantee the truth of a conclusion if validity itself depends on that self-same truth [Etchemendy, 2008, 265–6].

Now Carnap feigned to relinquish an epistemic conception of consequence with the principle of Tolerance, but his contrived selection of examples of languages with transformation rules that are in fact at least *prima facie* intuitively valid masks the real consequences of this. To the extent that Carnap's position is plausible, it is because of the air of intuitive validity that continues to pervade the transformations licensed in Languages I and II. Tarski, on the other hand, gives us a true generality conception of consequence.

What are we to make of Tarski's progress from the derivational conception of consequence still present in CTFL to the model-theoretic conception? Neither conception is satisfying: the derivational conception isn't a conception at all, but simply offers lists of rules that display the feature we seek to understand, while the model-theoretic conception is plagued by just the problems Etchemendy stresses: the only case in which it offers security is the one in which, as stressed by Kreisel [Kreisel, 1967, 152ff] Gödel's completeness theorem assures us that the model theoretic conception doesn't declare valid any arguments that the derivational conception doesn't. In other cases there is no reason to think that the model-theoretic conception even gets the extension of the consequence relation right, while the derivational conception manages this in hand-crafted cases but tells us nothing about what it is that makes certain rules good ones, nothing about how it is that we know that certain forms of inference preserve truth even when we are ignorant of the truth-values most of the substitution variants of the component sentences.

This fundamental question in the philosophy of logic remains open. What we can say in this historical study is that once Tarski discovered his way of reinstating the connection between consequence and truth while retaining the basics of Carnap's definition of L-consequence, he accepted the generality conception of logic that followed. As the identification of logic and mathematics this involved for STT was something he already accepted anyway, and since his extensionalism left him unperturbed by any apparent failure of the conception he offered to capture the apparent modality of the consequence relation, Tarski was perfectly happy with the result. And with this move there was nothing left of Intuitionistic Formalism, his adherence to which had already been weakened by his willingness to give up the theory of semantical categories. As we will see in the Conclusion, by the time of [Tarski, 1986e], all appeal to Intuition-istic Formalist notions of intuitive meaning, intuitively valid inference, linguistic conventions for the expression of thought, and so on, had been expunged from his thinking in favor of semantics, which now stood on its own.

8
Conclusion

8.1   Paris 1935 and the reception of semantics

When September 1935 rolled around, Tarski went out of his way to please his audience at the Unity of Science Congress in Paris. As we have noticed, he pitched the discussion of consequence largely as a modification of Carnap's notion of L-consequence, and he related misgivings he genuinely had about axiomatic semantics to a concern with physicalism that he probably didn't. Like Popper, who was made "joyful" [Popper, 1974, 398] by Tarski's views, Carnap had already been convinced [Carnap, 1963, 60], and Kokoszyńska was also present in support of Tarski's views. However, not everyone was enamored with "scientific semantics". Here is Carnap's recollection:

> When I met Tarski again in Vienna in the spring of 1935, I urged him to deliver a paper on semantics and on his definition of truth at the International Congress for Scientific Philosophy to be held in Paris in September. I told him that all those interested in scientific philosophy and the analysis of language would welcome this new instrument with enthusiasm, and would be eager to apply it in their own philosophical work. But Tarski was very skeptical. He thought that most philosophers, even those working in modern logic, would be not only indifferent, but hostile to the explication of the concept of truth. I promised to emphasize the importance of semantics in my paper and in the discussion at the Congress, and he agreed to present the suggested paper.
>
> At the Congress it became clear from the reactions to the papers delivered by Tarski and myself that Tarski's skeptical predictions had been right. To my surprise, there was vehement opposition even on the side of our philosophical friends. [Carnap, 1963, 60].

Additional discussion was arranged, and there a debate that was to continue for several years, in print and in correspondence, was carried out among Tarski, Carnap and Kokoszyńska on the one side, and Neurath, Naess and comrades on the other.

As [Mancosu, 2008] provides an excellent discussion of the ensuing debate, I will only indicate it here in outline. Neurath, who had already long objected to the idea that sentences are ever compared with reality rather than other sentences, and who had advocated replacement of "true" with "accepted" (in particular "in the Encyclopeida" of unified science) [Mancosu, 2008, 201], was adamantly opposed. Confronted with Tarski's proposal, Neurath had little to say on its technical merits but rather seems over a period of years relentlessly to have pounded the table over its potential to mislead the uninitiated into thinking there was such a thing as a comparison of a statement with reality. Kokoszyńska had taken Neurath to task over the obvious incompatibility between the identification of "true" with "accepted" and the T-sentences [Mancosu, 2008, 212] and, tellingly, Neurath at one point in a 1937 debate with Carnap suggested that the conception wasn't thus incompatible, since, taking "the Encyclopedia" to be the set of accepted sentences, one could adopt the usage of "is true" on which "'p' is true" is accepted into the Encyclopedia if and only if "p" is [Mancosu, 2008, 216]. Neurath's confusion is the common arising from the fact that anyone who defends an analysis of truth will, if they remain consistent, accept sentences if and only if these sentences have the property which by their account reduces the property of truth. This doesn't show that any analysis is comptible with the T-sentences—that is, with the truth conditions that sentences actually have—any more than the fact that the defender of an ethical theory with repugnant consequences can remain consistent by acting on these consequences shows that all ethical theories are compatible with what is actually right.[1] Sympathetic interpreters like [Mormann, 1999] tie Neurath's reaction into his overall set of views, but for our purposes this is the fundamental confusion. Carnap himself grew frustrated with Neurath over the years, writing to Neurath in 1943 that as early as 1935 "there were no rational arguments left on your side" [Mancosu, 2008, 220].

Tarski, on the other hand, seems to have been convinced by the episode that the long-standing policy of the Warsaw logicians dictating professional silence on philosophical matters was the right one [Murawski, 2004, 329]. Though as detailed in the set of conference notes toward the end of v. 4 of his *Collected Works* Tarski continued sporadically to participate in some philosophical events, his publications ceased to have even the modest philosophical cast of his early work; e.g. [Tarski,

1986d] simply presents the standard limitative results in a recognizably contemporary form without any hint of the Intuitionistic Formalist aspects of the original discussion in CTFL. The late "Truth and Proof", published in *Scientific American*, confirms this; clearly a set of mothballed remarks from the 1930s, the article shows that, to the extent that he had given the topics any thought at all, his views had not changed.

The one exception here is, of course, Tarski's most widely-read publication, 1944's "The Semantic Conception of Truth and the Foundations of Semantics". Tarski introduces the topic as giving a "satisfactory definition", that is, a "*materially adequate* and *formally correct*" one, of the "notion" of truth [Tarski, 1986e, 665]. Interestingly, note 4 notes that "notion" and "concept" can mean anything from a term itself, to "what is meant" by it, to its denotation (a usage found in [Tarski, 1986j]), but then waves off the issue with the comment that Tarski "share[s] the tendency to avoid these words in any exact discussion" [Tarski, 1986e, 694]. Tarski's previous interests in concepts and capturing them is clearly gone; what we have left is direct interest in semantics in its own right.

Tarski then announces that the aim of the definition is to "catch hold of the meaning of an old notion", notes that truth applies to sentences and is thereby relative to a language (with no more explicit concern for how this relativization is compatible with there being a single concept than he showed in the 1930s) and repeats, with some elaboration, the remarks on the "semantical definition" from [Tarski, 1983a, 155], this time calling the conception the "classical Aristotelian" conception and noting its imprecision. With no more explanation than in [Tarski, 1983a], capturing the meaning of "true" according to the classical Aristotelian conception is then equated with implying the T-sentences; the idea is that "if we base ourselves on the classical conception" we will say that "snow is white" is true if snow is white, and false if it isn't. Convention T is then introduced but left unnamed. Another central bit of Intuitionistic Formalism is therefore missing from SCT: the conception of language in terms of conventions. The requirement that the T-sentences be implied by the definition is now simply accepted in its own right, rather than being underwritten by a theory of the capture of concepts in the usuage of a term as determined by a definition or any other sort of convention.

Semantics is discussed as in ESS and the usual points are entered about an exact study being focused on formalized languages. After that Tarski rehearses Lukasiewicz's presentation of the liar paradox and then remarks, following [Tarski, 1983a, 157–8] that it shows that "semantically closed languages" must be left out of consideration. Tarski notes in a way consonant with [Tarski, 1983a, 165] that "everyday language" seems to be the sort of language that is to be disregarded and follows CTFL in attributing inconsistency to such language.[2] Tarski then explains the distinction between object language and metalanguage and notes that an adequate definition can only be formulated in an "essentially richer" metalanguage. This is then followed up by a thumbnail sketch of recursion on satisfaction and the definition of truth as satisfaction by everything. Tarski then asserts the "material" adequacy of the definition, and notes that "we can show that one and only one of any two contradictory sentences is true". As in CTFL, this is treated as a requirement of the Aristotelian conception once it is mentioned even though it wasn't built into the earlier definition of material adequacy. I take the repetition of this inconsistent treatment of "material" adequacy to be confirmation of the extent to which Tarski is running on auto-pilot in summarizing CTFL. In SCT Tarski is trying to get in the bare minimum of semantics without the Intuitionistic Formalist background from which it originally sprang, but he doesn't otherwise appear to have been thinking about the topics in the intervening years.

Most importantly, missing from SCT is any conception of meaning other than that provided by semantics. In note 20 to [Tarski, 1986e, 678] Tarski says that all of the semantic concepts mentioned in the section (§ 13) can be defined in terms of satisfaction, notes that consequence can be defined in terms of it as in [Tarski, 2002] and then, citing [Carnap, 1942, 55] mentions that synonymy can also be defined in terms of satisfaction. Since throughout CTFL "translates" and "means the same as" are used interchangeably, this is a significant change. While in CTFL translation was tied to the expression of intuitive meaning, in SCT it is an ordinary semantic concept like any other; in accord with Carnap's definition, two expressions are synonymous just in case they have the same extensional semantic value. Intuitionistic Formalism has now completely been abandoned as a conception of meaning in favor of semantics itself.[3] Having seen that the last piece of Intuitionistic Formalism that he accepted, the epistemic conception of consequence embodied in the definition of consequence in terms of stated rules of derivation, could go, Tarski shifted to understanding meaning in general in terms of the semantics he had introduced.

Tarski then turns to his "Polemical Remarks" in the second part of the article. His § 14 disclaims any intention to adjudicate between the classical conception of truth and various competitors, such as "the pragmatic conception, the coherence theory, etc." [Tarski, 1986e, 680]. Though officially this agnosticism and the concomitant thesis of the ambiguity of the everyday "true" among these notions figured already at [Tarski, 1983a,

153] one somehow gets the feeling that Tarski's claim here isn't sincere, especially given his scornful remarks on denials of the T-sentences later in the article. Then §§ 15–16 dispense with some confusions about the sentential connectives and the suggestion that Tarski's theory is a form of what Kotarbiśnki called "nihilism"; Tarski makes the obvious points and refers the reader to [Kokoszyńska, 1936].

Tarski then turns to the question of whether the semantic conception really offers a "precise form" of the classical conception. As noted above (§ 4.1.4), this is a good question. I have maintained that Tarski has, in principle, an answer when the classical conception is stated in the form of CTFL's "semantical definition", while, by contrast, there is good reason to think that the semantic conception isn't in any significant sense a correspondence theory (§ 4.3.4). More interesting, though, in addition to the rather plain inconsistency of Tarski's insistence here that the semantic conception does "conform to a very considerable extent with common-sense usage" with § 14, is Tarski's discussion of Arne Naess's early exercise in "experimental philosophy". Naess's survey showed that only 15% of people agreed that "true" means "agreeing with reality", while 90% assented to a typical T-sentence. Tarski "waives" consideration of whether the two conceptions are the same, though surely the result indicates just the disconnect between correspondence and the T-sentences we have noted. The conclusion I would draw here is that Tarski never questioned whether Kotarbiński had things right, and always concerned himself rather with how to express the conception in a deductive theory.

The beginning of § 18 is interesting in that Tarski seems to disclaim the philosophical relevance of his work that he stressed in CTFL: "in general, I do not believe that there is such a thing as 'the philosophical problem of truth'". Tarski had had it with philosophical controversy after the discussions following the Paris conference, and the ensuing pages manifest his frustration. Rather than continuing the philosophical project, Tarski in SCT presents his work on truth as of formal interest. Tarski makes the perfectly correct, though often misunderstood point that it is part of the semantic conception to equate the assertability conditions of "'p' is true" and "p", but notes that this says nothing about the conditions under which either can be asserted. He stresses the compatibility of the conception with any "epistemological attitude", though he also scorns any conception of truth itself that denies the T-sentences. The apparent inconsistency of this with § 14 is again left unexplained.

Moving on, § 19 concerns itself with Neurath's charge that the semantic conception is "metaphysical". Tarski notes that "metaphysical" in the objector's usage seems to be nothing more than an empty term of abuse, suggests several readings of "metaphysical", and argues that on all of the readings on which it expresses something bad, the semantic conception isn't "metaphysical". In particular, Tarski defends the incompatibility of his definitions with the verificationist requirement that a definition of a term give criteria for its applicability [Tarski, 1986e, 687–8]. Then § 20 discusses the role of semantic study in empirical science, § 21 ascribes a limited role to semantics in the methodology of empirical science, and § 22 discusses its greater role in metamathematics, and the way in which metamathematics, via arithmetization, is really a branch of mathematics. Finally § 23 closes by rejecting the idea that the value of a body of research is to be judged by its applicability.

In light of its haphazard composition and in particular the blatant inconsistency of § 14 with § 18—an inconsistency that could only be eliminated by taking seriously Tarski's feigned adoption of the view that anyone who disagreed with him about truth had changed the topic— SCT can only be read as an intended final attempt to get philosophers to stop bothering him. One can hardly blame him: though more cogent objections have played a significant role in the contemporary literature, the objections surveyed in SCT scarcely merit attention.

8.2   Final remarks

From this point one could discuss the important role Tarski's work played in the development of analytical philosophy, but since this story, unlike the one I have told, is relatively well known, our study comes to an end here. Tarski's philosophical career ended with something of a whimper: the conception that got him started, Intuitionistic Formalism, contained the seeds of its own destruction when Tarski turned it on semantic concepts. His approach to truth was greeted initially with incomprehension by its critics and to some extent its supporters. The techniques he introduced went on to an illustrious life in logic and the philosophy of language, but he himself went on in logic and mathematics; for an overview of his career one may consult [Vaught, 1986] and [Feferman and Feferman, 2004].

Though the conception of consequence is problematic, truth-theoretic semantics has many advantages over the Intuitionistic Formalist account, especially when it is detached from Tarski's definitional project and treated as issuing in axiomatic theories that state the truth conditions of the sentences of a language as recommended in [Davidson, 1990] and [Etchemendy, 1988]. In the place of hand-waving about the association of complex thoughts or thought-constitutents with complex expressions one has a straightforward, genuine account of semantic compositionality in Tarski's recursion on satisfaction. In the account of meaning this can only be regarded as enormous progress, although the problems of moving beyond the simple stock of modes of composition extensional methods can handle are well known.

As maintained in the introduction, neither the representational nor the expressive conceptions of meaning stands alone in the overall account of language and communication. Truth-theoretic semantics abstracts meaning from the attitudes of speakers in a helpful way, but it remains the case that expressions have the truth-theoretic properties that they do because of the attitudes toward them and use made of them by speakers in their environments. Building on Tarski's contributions we have, as we did not before them, a clear and workable conception of what it is that these factors determine when they determine meaning.

Notes

Introduction

1. One important connection to this mathematical work needs to be noted: Tarski's work in the period was, on many views, motivated in part by the desire to give rigor to the notion of a definable set. I discuss this matter to some extent in 3.3.2 below, but see also [Feferman, 2008], as well as [Vaught, 1974] and [Vaught, 1986] as well as works referred to therein.
2. For surveys of the mathematics, see [Duda, 2004], [Blok and Pigozzi, 1988], [Doner and Hodges, 1988], [van den Dries, 1988], [Levy, 1988], [McNulty, 1986], [Monk, 1986].
3. For comments on the manuscript I thank Greg Frost-Arnold, Wilfrid Hodges and Terje Lohndal. I also thank audiences at several presentations of material from the book: at the meeting of the Society for Exact Philosophy in Kansas City, in particular Stewart Shapiro, Philip Kremer, Michael Glanzberg and Paul Pietroski; at the School for Advanced Study at the University of London, in particular Barry Smith, Ian Rumfitt and Oystein Linnebo; at the University of Paderborn, in particular my host Volker Peckhaus; at the University of Albany, in particular Bradley Armour-Garb; at the Ludwig-Maximillian University in Munich, in particular Hannes Leitgeb, Jeffrey Ketland, Olivier Roy and Florian Steinberger; at the Truth at Work conference in Paris, in particular Henri Galinon and his fellow hosts, as well as Stephen Read, Gila Sher, Volker Halbach and Michael Glanzberg; and at the University of Siegen, in particular my host Richard Schantz. This book was written in part during the term of a *Forschungsstipendium für Erfahrene Wissenschaftler* granted by the Alexander von Humboldt Foundation for study at the University of Leipzig. I thank the foundation and especially my kind host, Professor Dr. Pirmin Stekeler-Weithofer.

1 Intuitionistic Formalism

1. One can already find the view in articulated well in Pasch, e.g. [Pasch, 1912, 98]. Detlefsen [Detlefsen, 2008] traces the fundamental ideas back much further.
2. One Warsaw figure who was a formalist in Hilbert's mode rather than Leśniewski's was the one who spent time in Göttingen: the Ajdukiewicz of the habilitation dissertation, who relegates the appeal to intuitive meaning to the primitive state of a deductive science, while in a mature axiomatization being meaningful is a matter of a sign's role in a theory, rather than some external relation of denotation [Ajdukiewicz, 1966, 4–6].
3. See [Betti, 2004, 278] for a discussion of the issues involved in interpreting Leśniewski's later work in terms of the early work.

4. Leśniewski's "adequately" (*adekwatności*) here anticipates Tarski's own notion of intuitive adequacy in a theory.

5. On Twardowski, founder of the Lvov–Warsaw School and Leśniewski's mentor, see [Woleński, 1989] and [Smith, 1994].

6. As Woleński notes [Woleński, 1989, 252], a Husserlian theory of meaning intention was fairly universally accepted in the Lvov–Warsaw School. For one comparative discussion concerning Husserl, Meinong and the early Russell, with some reference also to the British Empiricists, see the translator's introduction to Twardowski [1977].

7. Indeed, as we will see in § 2.2.1, at least in Tarski's hands the view is incompatible with compositionality.

8. [Betti, 2004, 269ff] discusses the traditional logical framework of Leśniewski's early papers.

9. The quotation is preceded immediately by the rather incoherent claim that Leśniewski doesn't "assert" his axioms but only that the theorems follow from them; this hiccup of commitment to "if-then-ism" about his deductive theories seems not to have survived the transition to the later work.

10. On the roots of Leśniewski's attitude here in Twardowski's opposition to "symbolomania", see [Smith, 1994, 182–3].

11. Carnap notes the influence of Frege on Leśniewski at [Carnap, 1963, 31]. The comparison between Frege and Leśniewski would repay more attention than I have been able to give it here.

12. See [Smith, 1994, 184] for a similar discussion of Leśniewski with commentary on the relation between his views and Twardowski's.

13. Hence Leśniewski's peculiar sentential connectives [Leśniewski, 1992f, 669–71]. The graphic design means that each connective actually is a display of its own truth-table, and that simple mereological relationships among the connectives display relations of implication and incompatibility among sentences formed with them. This logically useless innovation embeds reminders of the meanings of the connectives and the inferential relations thereby determined within every formula. This is otiose with respect to the logic itself, but is quite pertinent if what one is interested in is ensuring successful communication.

14. Leśniewski notes his own affinity for Kotarbiński at [Leśniewski, 1992h, 372].

15. See [Rojszczak, 2002, 45–6] on convention in the Brentano school, especially in Kotarbinski and Marty.

16. One finds the view also at [Ajdukiewicz, 1978a, 44ff].

17. The view, then, isn't the view criticized in van Orman Quine [1966], since what is held to be conventional is a function from expressions to thoughts and thought constituents; the meanings and logical relations among the latter are held by neither Kotarbiński nor the later Leśniewski to be somehow "assigned" to them, in turn, by convention.

18. One might wonder here how this appeal to types is to be reconciled with Kotarbiński's nominalism, but for our purposes it will suffice to note that Kotarbiński, for whatever reason, has no particular qualms about speaking of "equiform" sentences [Kotarbiński, 1966, 6].

19. On the Brentanian Background, see [Woleński, 1989], [Woleński and Simons, 1989], and [Smith, 1994].

20. The view survives into the later work on actions and products; see e.g. [Smith, 1994, 177–8] where Twardowski's 1912 view is still that the significance of a sign lies in its power to cause ideas in people's minds; as Smith there writes, "Communication and mutual understanding are possible on this account, not because our words or sentences relate to Platonic meaning-entities capable of being entertained simultaneously by different subjects, but because our words are able to evoke in others mental processes which are in relevant respects similar to those mental processes which they were used to express".

21. One finds this sort of view in Ajdukiewicz, too: e.g. [Ajdukiewicz, 1978b, 6]: "Speaking Polish ...consists not only in producing speech-sounds which belong to Polish phonology but also in experiencing, simultaneously and in connection with those speech-sounds, mental acts in a manner assigned, whether uniquely or ambiguously, by the rules of Polish to those speech sounds". Ajdukiewicz differs from Kotarbiński in finding, *via* Husserl [Ajdukiewicz, 1978a, 37] a tighter connection between language and thought, a view that was a central aspect of his "radical conventionalism", though like Kotarbiński he rejects simplistic associationist accounts of the relations between thought and language [Ajdukiewicz, 1978b, 7–20]

22. This point of view was common in the Peano school; see e.g. [Pieri, 1901].

23. The passage is from 1937, but given its clear connection to Padoa, Leśniewski and Kotarbiński, we can feel safe quoting it here as representative of Tarski's views some years earlier.

24. [Mancosu et al., 2009, 323] agree that the notion of interpretation in Padoa isn't the current semantic one and give a reading in terms of sentences with meaning that is largely consonant with the one here.

25. Leśniewski also makes reference to the "well-known method of 'interpretation'" at [Leśniewski, 1992h, 261].

26. As [Murawski, 2004, 330] notes, following Twardowski members of the Lvov–Warsaw school were opposed to "psychologism" but what they were opposed to was psychologism *about logic*, the view famously criticized by Frege, on which logical laws are empirical generalizations about how people actually think. The psychologism at issue here is psychologism about semantics—the far more plausible doctrine that meaning is a matter of the attitudes of speakers toward symbols. See also [Woleński, 1989, 262–3].

27. See [Ferreirós, 1999, 348ff] and [Reck, 2004, 164ff] for discussion of the simple theory of types and developments in type theory in the 1920s.

28. The distance travelled from Twardowski, who is quite happy to compare contents of "presentations" (*Vorstellungen*) to paintings of landscapes [Twardowski, 1977, 11–17], is perhaps significant, then, though it goes by subtraction.

29. Similar remarks persist into the textbook, e.g. [Tarski, 1995, 118].

30. One readily associates the view rejected here with Hilbert, but Tarski probably also has Ajdukiewicz in mind, e.g. [Ajdukiewicz, 1966, 14]: "Since no such element with intuitive meaning is a component of formalized axioms, none of them may be regarded as a sentence in the intuitive sense ... Since axioms are not sentences in the ordinary sense, and in their ordinary meanings [the] words 'true' and 'false' refer to sentences exclusively ... axioms cannot be judged from this point of view".

31. Interpreters who agree that the notion of interpretation at issue in this passage is roughly the one I have suggested include [Raatikainen, 2003, 41ff], [Fernández-Moreno, 1997] and [Milne, 1997]. [Hodges, 1986, 147ff] offers the more standard interpretation of these remarks because Hodges takes the only possible notion of interpretation to be the model-theoretic one.

32. One finds a similar view of connotation (also with reference to Mill) and its capture in the inferential structure of a deductive theory at [Ajdukiewicz, 1978b, 21]: "A term N connotes in English properties $C_1, \ldots, C_n$ if and only if every English-speaking person is prepared, on the basis of his belief that an object A has properties $C_1, \ldots, C_n$ to accept as an immediate conclusion an singular sentence of English whose subject is a name of A, whose predicate is N and whose joining copula is the word 'is'; and vice versa, on the basis of his acceptance of such a singular sentence he is prepared to accept that A has the properties $C_1, \ldots, C_n$". Ajdukiewicz goes on to reject this view (due to perceived problems with the extensionalism about properties he takes it to involve) and to replace it with an entirely intra-systemic conception of meaning of the sort that we have attributed above to Hilbert—compare, for instance, his definition of synonymy at [Ajdukiewicz, 1978b, 31-2] with Carnap's at [Carnap, 2002, 42]. Thus, not surprisingly, given that he was with Hilbert in Göttingen [Ajdukiewicz, 1978c, xiii] while Leśniewski was working with the view expressed at [Leśniewski, 1992b, 16-7], Ajdukiewicz is in 1931 a standard rather than an Intuitionistic Formalist; indeed, [Ajdukiewicz, 1978b, 33-4] explicitly notes the incompatibility of his conception of meaning with any non-trivial form of the thesis that the use of a sentence expresses a thought and suggests reducing our conception of a having a thought to a disposition to assent to a sentence. As we will note below § 6.2 there are therefore corresponding questions about what his conception has to do with intuitive thought. That Ajdukiewicz rejects the view is further evidence that it was there for Tarski to accept, though we might speculate that Tarski's own failure to explicate the notion of the content of a concept in terms of connotation conceived in a traditional way bore some significant relation to Ajdukiewicz's criticism.

2 Tarski as Intuitionistic Formalist

1. [Suppes, 1988, 81] attributes the first axiomatization of the notion of consequence to [Tarski, 1983g]. See [Wybraniec-Skardowska, 2004] for a discussion of Tarski's approach here and some later extensions of it.

2. See [Arpaia, 2006] for a historical discussion of Tarski's approach and its connections with algebraic logic.

3. See also [Tarski, 1983d, 39, 55] for simpler definitions that are still hardly "intuitive".

4. I am thus inclined to disagree with David's [2008] attribution of an implicit treatment of "true" as indexical in Tarski's metatheories.

5. Smith considers problems raised by indexicality for Twardowskian views about meaning and communication at [Smith, 1994, 180–1].

6. The idea that relativization also makes for "ambiguity" is found at [Tarski, 1983f, 110]. However, the ODS passage can be taken so that a metalinguistic

expression is "ambiguous" if its extension varies by the languages it is applied to, which is the view offered here.

7. In a fuller account we could trace the history of the two conceptions of definition further back and outside of Poland; but for two obvious statements of the abbreviative conception, see [Russell, 1903, 429] and [van Orman Quine, 1966, 71].

8. [Ajdukiewicz, 1978a, 54] sides with the theory-relative conception, though the earlier 1921 dissertation seems to endorse the abbreviative conception [Ajdukiewicz, 1966, 16 n. 2].

9. At one point Kotarbiński compares axiomatic "pseudo-definitions" to learning language by the "Berlitz method". The comparison fails, since what might be difficult about learning the meaning of a primitive expression without metalinguistic commentary does not carry over to learning a defined one. When it comes to primitives, even Leśniewski agrees that they need metalinguistic commentary.

10. Tarski's account appears to have enjoyed some success back in Warsaw; notes on Leśniewski's lectures from the 1933–34 term include a version of Tarski's account [Leśniewski, 1988, 168].

11. Here I disagree with other readings of Padoa, e.g. [Suppes, 1988, 84] and [Hodges, 2008, 105ff].

12. Note here that Kotarbiński's own explanation of Padoa's method is transparently psychologistic, in that giving an interpretation is "attaching the meaning" of a specified term to a symbol, and for Kotarbiński the meaning of a term is the content of a concept [Kotarbiński, 1966, 255]. Likewise, Ajdukiewicz's 1921 discussion of proofs of consistency of axiom systems by the giving of examples, with reference to Pieri, takes an "example" to be a set of interpreted sentences so that, as Ajdukiewicz construes it, "giving an example" for an axiom system isn't sufficient for proving that it is consistent without a guarantee that the example itself isn't "self-contradictory" [Ajdukiewicz, 1966, 22]. Nobody thinking in model-theoretic terms would say this. Ajdukiewicz ascribes "intuitive meaning" [Ajdukiewicz, 1966, 6] to sentences, but doesn't link this to anything model-theoretic, either. (Waters are muddied here by the Husserelian and Meinongian § 3 of Ajdukiewicz [1966] with its inconsistent objects, etc. This would bear attention, but not here.)

13. Hodges reads Padoa as offering an implementation of his "method" in terms of a notion of a model which I simply cannot find in [Padoa, 1967], offering the justification that this is "the most straightforward reading of Padoa's claim" [Hodges, 2008, 105]. He then notes that Tarski's procedure is essentially more complicated than Padoa's *so understood* and is restricted by features of simple type theory on which Tarski relies. This is quite right about Padoa as Hodges takes him, but depends on ignoring Padoa's essential agreement with Tarski that an "interpretation" is an idea mentally associated with a symbol.

14. A fair amount of ink, a good deal of it critical (e.g. [Devidi and Solomon, 1999]), has been spilled over what Tarski means by an "essentially richer" language, but we are told exactly what it is here. See [Rogers, 1963, 28] for an early discussion of the notion, and [Ray, 2005] in response to Devidi and Solomon.

15. A comparison to Carnap's work on "General Axiomatics" in the late 1920s might be fruitful here, since Carnap was interested in "monomorphism", that is, categoricity (indeed, Tarski uses both terms) but failed to distinguish syntax from semantics or object- from metalanguage [Reck, 2004, 169]; since he held in the same period that *Bedeutung* is a *Scheinrelation*, his conception may well have been close to the one Tarski offers.

16. Tarski can still account for the apparent difference between "human" and "featherless biped" in a theory that implies their extensional equivalence: the latter, but not the former, has parts that bear the intuitive meanings of featherlessness and bipedality. In the terms of [Douglas Patterson, 2005, 326–52] the theory is neither compositional nor reverse compositional.

17. A different issue is what Tarski would or should say about extensions of a theory that involve not just new expressions, but new contexts—e.g. modal or attitude contexts. In Patterson [2008b] I assume that he would hold the the view that two terms equivalent in an extensional theory might nevertheless have meanings that would distinguish them in an intensional theory. This is, roughly, the analogue of the view that theories admitting of this possibility aren't monotransformable: there are extensions of the theories that distinguish expressions that are definable in terms of one another in the theories themselves. Since Tarski certainly holds that good theories can fail to be monotransformable, he might accept the analogous claim about his own metatheories once intensional contexts were in view, but now that I have a better grasp on what is going on in Tarski [1983k] I am a bit less confident of this.

18. Adding axioms to this effect is, historically, acquired from Langford, who also includes such axioms in the theories he investigates (cf. [Scanlan, 2003, 318], [Mancosu, 2006, 221]).

19. Cf. [Mancosu, 2006, 222–3]. The distinction between intrinsic and absolute categoricity, as Mancosu notes, is important to assessing the claims of some interpreters. e.g. Gómez-Torrente [1996], that Tarski's use of domain-specifying predicates shows that he had a variable domain conception of a model. See § 7.3.1 below.

20. There is a worthwhile comparison here to Ajudkiewicz's 1934 remarks on "open" and "closed" languages. For Ajdukiewicz a closed language is one such that a new expression added to it with a meaning not shared with any of the old expressions produces a language that is "disconnected" in the sense that the meaning of the new expression bears no constitutive relationships with the meanings of the old ones; by contrast, an open language can have an expression added to it whose meaning is in part dependent on the meanings of old expressions [Ajdukiewicz, 1978a, 50–52]. Theories complete with respect to their specific terms correspond to Ajdukiewicz's closed languages, and the point that a language is complete with respect to its specific terms just in case it has no categorical essentially richer extensions then corresponds to Ajdukiewicz's thesis that a term with a new meaning added to a closed language produces a language that is disconnected [Ajdukiewicz, 1978a, 51]. The importance of completeness so conceived then corresponds to Ajdukiewicz's thesis that closed languages completely determine meaning in the sense that any two closed languages that share an open part perfectly translate one another.

21. More could be done than has been done here to unearth the sources of Tarski's remarks on categoricity, especially given the discussion of geometry and mechanics that closes the article. Tarski cites Fraenkel [?, 347–54] on categoricity [Tarski, 1983i, 390], but the relevant passages in Fraenkel serve merely to disambiguate categoricity from other meanings of *"vollstandigkeit"* current at the time, though Fraenkel's references to Mach on the unity of science indicate that something more philosophical is in the air as well. One avenue of exploration would be the connection between this section of Tarski [1983k] and Carnap's work in the 1920s on axiomatizations for space-time, since Carnap from the dissertation of 1921 through the decade singles out *"Eindeutigkeit"* as an important feature of an axiom system, a mark of its having captured reality, and the term and its cognates appear in Tarski's German. For a start on this topic, see Howard [1996] and [Awodey and Reck, 2002, 25–7]. The connection to Hilbert's various "axioms of completeness" would also merit further study, though as noted at [Awodey and Reck, 2002, 12–14] though the axioms imply categoricity they themselves are not metalinguistic (a point also made by Fraenkel [Fraenkel, 1928, 347]).

22. I take myself, then, to provide some explanation for a fact that Hodges notes [Hodges, 2008, 113], that Tarski never claims that the terms in favor of which a defined term is eliminated are somehow clearer or more basic than the defined term itself.

23. One small bit of evidence that ideas like this were in the air in Warsaw: a critic, discussing Łukasiewicz's mania for shortening his axioms, writes that "Usually the shortest ones ... have no application in practice ... Łukasiewicz was even a theorist of that sport; he claimed that the intuitive content of an axiom system is not important, since it is only the set of those theorems which are its consequences that count" [Woleński, 1989, 91].

3 Semantics

1. For a discussion of the main line of development of thought about quantification along semantic lines, by contrast, see [Goldfarb, 1979].

2. Compare here the discussion of the correspondence between Gödel and Carnap in 1932 where Gödel points out a manifestation of this problem in a particular case at [Coffa, 1987, 552–3]. (For a proposed work-around along Carnap's pre-1934 lines, see [de Rouilhan, 2009, 139–40].)

3. Setting aside Leśniewski's putative alternative discussed below.

4. See [Simons, 1993, 215ff] for one discussion.

5. By Zermelo, as noted by Fraenkel at [Fraenkel, 1928, 210]

6. The distinction is often erroneously credited to Ramsey. See e.g. [Ferreirós, 1999, 351].

7. The peculiar suggestion that the paradoxes have something to do with intensionality is bound up with Carnap's understanding of the latter. The *Abriss* refers the reader to §§ 43-45 of the *Aufbau* on this point.

8. For a more thorough history, see [Mancosu et al., 2009, 419ff].

9. See [Mancosu et al., 2009, 326–8].

10. [Fraenkel, 1928], which Tarski cites, is also completely semantic; see, e.g., the discussion of categoricity at 349ff.

11. Sometimes the inspiration for Tarski's development of semantics, in particular the use of the notion of satisfaction, is attributed to Ajdukiewicz's 1921 habilitation dissertation (e.g. [Betti, 2008, 61ff], [Mancosu et al., 2009, 435]). The attribution is problematic, however; consider the treatment of an "example" intended to show a set of axioms to be consistent as being a set of sentences that itself might be inconsistent [Ajdukiewicz, 1966, 22, 32]. No model in the semantic sense can be "inconsistent". Things are complicated here, however, by the Husserlian and Meinongian section 3 of the work, in which Ajdukiewicz maintains that "inconsistent objects" exist and are "created by our minds". Clearer, perhaps, is Ajdukiewicz's discussion of truth and satisfaction in that section, but in this we get something that is, though clear, certainly not semantic, namely "the definitions: 'Object P satisfies the propositional function $f(x)$ means — the [constructing] definition of P entails $f(P)$' and 'the sentence $f(P)$ is true means — there exists an object P satisfying $f(x)$'." Satisfaction here, such as it is, is an entirely intra-linguistic matter of a certain privileged definition of a term implying the closure of an open sentence by that term ([Betti, 2008, 63] agrees). By this definition, the encouraging dictum "the domain of a theory is the set of all objects satisfying the axioms" [Ajdukiewicz, 1966, 41] can't be read semantically, since the effect of the definition of satisfaction is to equate the apparently semantic "satisfies" with the existence of certain implications among sentences of a theory. Perhaps these passages got Tarski thinking, but they are so confused—note the rampant use–mention sloppiness and the insouciant phenomenological doctrines that lead to the view that there are "objects" "in the domain" of a theory that neither exist nor have definitions consistent with its axioms [Ajdukiewicz, 1966, 43]—and are so far from clearly semantic that little anticipation of Tarski's semantics can be attributed to them.

12. [Heck, 1997, 545–6] understates the importance of explicit definition for guaranteeing consistency, taking the distinction between object and metalanguage to be sufficient. But the problem with this is that it isn't always obvious when we have a distinction between object and metalanguage in the relevant sense. Ascending to a metalanguage blocks one known sufficient condition for inconsistency, but as long as the theory remains axiomatic the worry about inconsistency remains. Only explicit definition guarantees (relative) consistency, and it was for this that Tarski was looking.

13. See [Addison, 2004] for more historical background and a survey of later developments of the mathematical aspects of the article.

14. See [Hodges, 2008] for a good discussion of the mathematical developments here and their relation to Tarski's work on quantifier elimination.

15. On a related note Feferman, taking himself to disagree with the claim of [Hodges, 1986] claims that "the notion of truth in a structure is present implicitly" the article [Feferman, 2008, 80–2], but, like the claim of [Hodges, 2008] Feferman's reading suffers from following Tarski in taking semantic definition of the set of things definable in L for a formal definition of "defines in L".

16. Cf. [Hodges, 2008, 111–12], who notes correctly that the passage is odd in that what is to be shown isn't even expressed in the theory.

17. "The empirical method" is a phrase Tarski uses for omitting a proof by induction and simply considering a few cases [Hodges, 2008, 98].

18. [Sinaceur, 2001, 54] notes Tarski's claims about what he has accomplished but does not comment on the distance between these claims and what Tarski actually does. See [Vaught, 1986, 871] for a positive take on this as mathematics.

4   Truth

1. [Ajdukiewicz, 1966, 21] calls a theory "adequate" (*"adekwatny"*) if its axioms, when given an intuitive interpretation, suffice to prove all intuitive truths expressible in the thereby interpreted language. Tarski's usage bears some resemblance to this, as well as to Leśniewski's use of the same Polish cognate for a notion closer to Tarski's at [Leśniewski, 1992b, 16–17].
2. I couldn't find a definitive source on this, but since Tarski otherwise wrote his own German (as his son is reported as saying in the introduction to [Tarski, 1992]) it seems more likely that he wrote the Postscipt in German himself than that he wrote it in Polish and had Blaustein translate it.
3. The classic on the tradition leading up to Tarski is [Woleński and Simons, 1989]. Some useful sources on the general situation before Tarski include [Niiniluoto, 1999b] on the situation in the German-speaking world in the 1910s and 1920s, and [Sluga, 1999] on Frege, Russell, Wittgenstein and their influence on Carnap in his reception of Tarski (although Sluga's reading suffers somewhat from the confusion of Tarski with Davidson, and of Tarski's "theory" with a Tarskian definition (§ 4.3.2)).
4. I leave for later the question as to whether it really is a correspondence theory.
5. On whom see [Woleński, 1989].
6. Advocating redundancy would seem to have been apostatical in the Lvov–Warsaw school, since Twardowski himself endorsed various correspondence formulations and, at [Twardowski, 1977, 96] has a discussion of "affirmations of affirmations" that, like Kotarbiński's discussion of "verbal" and "real" interpretations of "true", clearly distinguishes one sense in which they are redundant from another in which they are not.
7. Especially after the publication of [Tarski, 1986e] a significant debate developed as to whether the "classical conception" Tarski takes over from Kotarbiński as the "semantic conception" is really incompatible with the "utilitarian" conception. At least in the mid-1940s there was a significant movement in the literature to deny this—e.g. [Black, 1948, 61], [Perkins, 1952, 581ff]. This is subject to now-familiar counterexample, as Tarski himself explains at [Tarski, 1986e, 686]: if $\forall x, x$ is true iff x coheres with a set of beliefs or facilitates successful action when believed, we can't also hold that $\forall x, x$ is true iff $x = \ulcorner p \urcorner$ and p, since this gives us the obviously false claim that $\forall x$, accepting x facilitates successful action (etc.) iff $x = \ulcorner p \urcorner$ and p, since the psychological and other empirical facts may conspire to make false beliefs facilitate successful action (etc.) or true ones fail to do so. See [Haack, 1976b] for a more recent consideration of the issues, especially [Haack, 1976b, 240] where the supposed compatibility of "bizarre" theories of truth with the T-sentences is wrongly argued for by pointing out that anyone who thinks that all and only truths are asserted in the Bible will accept that "p" is asserted in the Bible iff he accepts that p, since, "if he is wise" doing the first will lead to

his doing the second; to say this is simply to point out that there is a valid argument from the claim that all and only truths are asserted in the Bible to the claim that some particular thing asserted in the Bible is true. [Keuth, 1978, 320] also advocates Haack's claim, and says that Tarski did, too. But this is to confuse the remarks in [Tarski, 1986e] on "epistemological attitudes" with the theory of truth.

8. I am aware that Tarski says that "In this construction I shall not make use of any semantical concept if I am not able previously to reduce it to other concepts" [Tarski, 1983a, 153]. But here, as in other passages, Tarski is being so loose with "concept" that he treats it as synonymous with "term": see the sentences that precede the one I have quoted. If a particular definition really reduces a semantical concept to other concepts, then the view that each definition introduces a different concept is correct. This would be incompatible with Tarski's constantly writing as though there is one and only one concept of truth. Given the tension, I diagnose verbal sloppiness rather than blatant inconsistency.

9. [Fine and McCarthy, 1984] discusses the extent to which the recursion on satisfaction is necessary, responding to questions raised in [Wallace, 1970], [Tharp, 1971] and [Kripke, 1976].

10. On this phrase, see § 3.3.2.

11. See [Woleński, 1989, 147–9] and [Ferreirós, 1999, 355–6].

12. We will see this issue again in § 7.4.1.

13. Here is a far from complete list: [Davidson, 1990, 295], [Etchemendy, 1988, 54], [Feferman and Feferman, 2004, 10], [Gómez-Torrente, 2004, 28], [Heck, 1997, 541–2], [Hodges, 2008, 115], [Ray, 1996, 625]. There are some exceptions in the literature, but not many: [Rogers, 1963, 26],ˈ[García-Carpintero, 1996, 114–5], [Mou, 2001, 103–4].

14. Something close to this practice can be found in Leśniewski, though in the first person singular, e.g. [Leśniewski, 1992e, 471].

15. Interestingly the opposite view is also attested: [Levison, 1965, 386] holds that Convention T states a necessary but not a sufficient condition.

16. A somewhat related error here is found in speculation sometimes heard as to whether or not the notion of consequence at issue in Convention T is stronger than material consequence. The question has a false presupposition, since the only notion of consequence Tarski has in CTFL is derivation according to stated rules.

17. For brevity I list some of the best-known references only.

18. Do not, of course, be distracted by the fact that a theory stated in that counterfactual situation would have needed to have "'snow is white' is true iff snow is white" as a theorem.

19. [Milne, 1999, 150] and [Ray, 1996, 662] beat me to the basic point, something I regret having overlooked there: a definition added to a logical theory isn't itself a logical truth.

20. "genuinely": non-disjunctive, stated in the vocabulary of such-and-such a physical theory, supporting counterfactuals, etc., as the reader pleases.

21. In case it needs mentioning, Field later abandoned these views for deflationism, in a way that amounts to accepting the criticism at [Leeds, 1978, 120–3].

22. [Cummins, 1975] remains an interesting early response to Field's conception of a physicalistically acceptable theory of reference that parallels developments in the philosophy of mind and philosophy of science concerning elimination and reduction in general. Soames [Soames, 1984, 419ff] and others point out that by Field's standards Tarski's recursion clauses are just as objectionable, since they too fail to support counterfactuals and so forth. [McDowell, 1978] accepts that Tarski's theory as he understood it isn't physicalist, as he understood that, but that it can be understood along broadly Davidsonian (e.g. [Davidson, 2002a], [Davidson, 1990, 299–300]) lines as instrumentalist about sub-sentential semantics and physicalist at the level of the confirmation of T-sentences. There is, in turn, a substantial literature on this sort of Davidsonian view; for a recent discussion in the context of interpreting Tarski, see [Hoffmann, 2007, 161–70]. [Kirkham, 1993, 289–302] argues that "physicalism" for Tarski, following Neurath and Carnap, required reduction to observables, not physics proper, and that Tarski succeeds in the project so conceived, though it is a bit puzzling how Kirkham takes it that a Tarskian definition reduces semantic notions to "observable" entities; the only secure contention in the article is that Carnap and Neurath, and therefore Tarski, required only that the "reduction" of semantics be extensionally adequate.

23. [Rojszczak, 2002, 32] suggests that Vienna Circle physicalism did not motivate Tarski's "choice of sentences for truth-bearers", but this question is narrower than the general question of whether physicalism motivates his attempt to eliminate semantic terms from his theories.

24. SCT in 1944 says that the dictum "The truth of a sentence consists in its agreement with (or correspondence to) reality" [Tarski, 1986e, 666–7] is a version of the classical Aristotelian theory which is also expressed by CTFL's "semantical definition".

25. See, e.g., [Hawthorne and Oppy, 1997] for a catalogue of possibilities.

26. The thesis that "'p' is true" is equivalent to "p"—something like the Redundancy theory for sentences, perhaps most familiar today from prosentential theories—was known in Poland at the time as "nihilism" about truth. Wolenski and Murawski attribute it to Zawirski and we have found it in [Czeżowski, 1918]. Kotarbiński discusses and rejects it in [Kotarbiński, 1966]. It was recognized at the time that Tarski's view was not a form of nihilism; see [Kokoszyńska, 1936, 162–5] who stresses the ineliminability of the Tarskian "true" in various contexts.

5 Indefinability and Inconsistency

1. [Hintikka and Sandu, 1999, 218–221] emphasizes this aspect of Tarski's views on indefinability.

2. Compare here [Tarski, 1986d], in which all of the Intuitionistic Formalist aspects of the discussion in CTFL are gone and we have in their place a clear, concise treatment of the results as they are now understood.

3. [Gómez-Torrente, 2004, 34] notes the point about the diagonal lemma and tries to write it off, but the reason given for attributing the diagonal lemma to Tarski's presentation—that (2) "applies to all negations of

predicates"—doesn't take sufficient account of the point that Tarski assumes that a class rather than a predicate is defined at the top of 250.

4. Moreover, nothing like the current version of the result, where Tarski's theorem about truth is distinguished from the more general result that no predicate of the language of a theory that extends Q can define the set of sentences that it semantically implies can be at issue in the passage, since in 1933 Tarski has no notion of consequence other than provability.

5. If I am not mistaken this is the only one-sentence paragraph in the work.

6. Unlike in the discussion at [Tarski, 1983a, 197] (§ 4.2.1) Tarski doesn't here specifically say that proving such theorems is required for intuitive adequacy.

7. Certainly available at the time, since [Ajdukiewicz, 1978a, 38] utilizes it.

8. My own approach to the paradox is along these lines. See [Patterson, 2008c], [Patterson, 2009].

9. Sometimes commentators [Levison, 1965], [Ray, 2003] try to dodge the straightforward reading of Tarski's claims by arguing that he holds that "colloquial" language is too ill-formed really to be inconsistent (or consistent). In addition for there being no need to absolve Tarski here, this strategy doesn't square with Tarski's willingness to call such language "inconsistent" [Sinisi, 1967].

10. The view I offered there was roughly that found at [Herzberger, 1970, 164], and it was mistaken as a reading of Tarski's notion of a "universal" language for the same reason, in that both readings take universality in terms of sets defined rather than concepts expressed. Martin's response to Herzberger [Martin, 1976] also retains the standard but misguided conception of what Tarski takes "universality" to be.

11. An interesting comparison: [Ajdukiewicz, 1978e, 81] denies the possibility of languages that are universal in Tarski's sense precisely because he lacks the Intuitionistic Formalist distinction between language and the thoughts expressed in it; his position corresponds to the one Tarski holds for formalized languages and, to some extent, to the position I attributed to Tarski in 2006.

6 Transitions: 1933–1935

1. The same point therefore goes for Tarski's remarks on axiomatic truth theory and its supplementation by the ω-rule.

2. Thus the whole point of the Postscript is that the difference between the theory of semantical categories and Carnap's theory of Levels matters, contrary to the conflation of the two at [Coffa, 1987, 560]. Coffa basically falls into the error of thinking that the *only* thing lacking in § 5 relative to the Postscript is the notion of an expression of infinite order [Coffa, 1987, 563]; note that Coffa's own explanation of Carnap's theory of levels at 551–2 doesn't mention the point that was important for Tarski, the provision that a functor could take arguments of variable level.

3. Again, these points aren't directly relevant to the body of the work, since they concern Carnap's theory of levels rather than STT.

4. Tarski concludes the postscript with an elegant (though non-constructive) presentation of Gödel's first incompleteness theorem; since the application is straightforward it need not concern us here.

5. This aspect of the relationship between Tarski and Carnap is much studied; see, e.g. [Woleński, 1999], [Coffa, 1987], [de Rouilhan, 2009].

6. A similar impetus to Tarski's development at the time may have come from an interchange with Ajdukiewicz, whose 1934 publication [Ajdukiewicz, 1978a] attempted, in a less sophisticated but strikingly similar way, to define meaning in purely syntactic terms [Ajdukiewicz, 1978c, 315]. Ajdukiewicz's view was subject to a simple counterexample by Tarski, who pointed out that it implied, for a theory consisting only of $a \neq b$ and $b \neq a$, that a and b were synonymous, despite their having, by the theory itself, different denotations. Though the work was not as sophisticated as *Logical Syntax* this surely provided a striking demonstration that syntax alone might not be trusted to deliver a good account of synonymy and other meaning-theoretic notions.

7   Logical Consequence

1. A note added to the English translation [Tarski, 1983g, 30], distinguishes the article as concerned with what is "derivable" and notes that what we get in [Tarski, 2002] is "semantic consequence".

2. Stroinská and Hitchcock retain "material adequacy" in their annotations, in accord with their stated aim of adhering to the standard English terminology [Tarski, 2002, 173]

3. As Tarski himself mentions in a footnote added later, the account is similar to the one offered by Bolzano in the *Wissenschaftslehre*. Since Bolzano had a significant influence on the Lvov–Warsaw school through Twardowski's early training [Smith, 1994, 157] and also though other routes (e.g. Husserl), one can't rule out some influence on Tarski here. On Bolzano, see [Lapointe, 2011].

4. [Kneale, 1961, 94–7] pointed out the overgeneration problem with respect to both Bolzano and Tarski long before Etchemendy. It wouldn't surprise me if the point had been made by someone about Bolzano even earlier.

5. [Mancosu, 2006, 215] suggests devoting a paper to the relationship between Carnap and Tarski on models and consequence. [Gómez-Torrente, 1996, 146] also notes the connection to Carnap, but it plays only a minor role in the paper's interpretation. By contrast, for example, [Sagüillo, 1997, 235–7] mentions "logicians" whose conception Tarski was trying to capture and discusses Russell, but neglects the article's overt discussion of Carnap.

6. Kotarbiński likewise identifies logic as the study of sentences in virtue of their logical form, where this is understood as being built out of only variables and logical constants, and then in turn simply lists the logical constants [Kotarbiński, 1966, 131], in his case the sentential connectives and the copula. Tarski still finds an enumerative definition of the logical constants acceptable in [White and Tarski, 1987].

7. Carnap himself is unable to avoid characterizing consequence in the usual way at [Carnap, 2002, 216].

8. For a systematic comparison of the Lvov–Warsaw school with the Vienna Circle, see [Woleński, 1989, 295ff].

9. [Hodges, 1986, 138] is often cited on this point, but note that Hodges doesn't explicitly propose the understanding of the article as a response to

the putative problem that logic should be free of commitments as to how many objects exist.

10. [Bays, 2001, 1710] argues that Tarski must have held a variable domain conception since he lets V be different in different applications of STT; Mancosu's response to this is correct: Tarski's conception is a "weak" fixed domain conception.

11. [Simons, 2009, 4] attributes Tarski's focus on absolute, non-model-relative truth to the influence of Twardowski's attack on relativism.

12. [Kreisel, 1967, 139] makes a different but somewhat related point about [Gödel, 1967a], focusing on Gödel's avoidance even of the idea of the general notion of logical truth as truth in all set-theoretic structures.

13. Or, alternatively, Hilbert and Ackermann would have asked a less general question and Gödel would have proved a more restricted result, commonly referred to as "weak" completeness [Etchemendy, 1988, 76].

14. Again, compare here [Łukasiewicz, 1970a, 17–18], which gives a definition much like Tarski's condition (F) and then happily accepts the result that various intuitively mathematical consequences are "logical".

15. I have neglected the corresponding problem of "undergeneration" in this section. Etchemendy is right, however, that it is also a threat: any restriction of $\mathfrak{F}$ intended to avoid overgeneration threatens to declare certain intuitively valid inferences invalid—e.g. taking "$=$" out of $\mathfrak{F}$ would rid the account of troublesome mathematical consequences for first-order logic, but wouldn't correspond to the conception of logic prevalent today. Another way of stressing undergeneration is found at [Hanson, 1999, 610]. Since the account requires models to be sets, consequences of theories with intended interpretations that aren't sets (e.g. set theory itself) won't be accounted consequences by the account. See also [McGee, 1992, 279].

16. [Jacquette, 1994] suggests not that Tarski's account quantifies over *possibilia*, but that the obvious falsehood of the Reduction Principle for ordinary extensional semantics motivates a "Meinongian" semantics that does so.

17. A more serious error comes in Ray's response to Etchemendy's claim that Tarski's notion doesn't capture the intuitive epistemic features of deductive consequence, the point being that, if all there is to logical consequence is what is captured by Tarski's definition, then "in general, it will be impossible to know both that an argument is 'valid' and that its premises are true, without *antecedently* knowing that its conclusion is true' [Etchemendy, 1990, 93]. Ray suggests in response that if φ follows from Γ by Tarski's account, and if all members of Γ are true, then it follows by Tarski's definition that φ is true. As Etchemendy and others have pointed out, this makes the mistake of using a definition as a premise in an argument for its own adequacy.

18. That the consequences of true sentences being true is an important fact about consequence is something Tarski's commitment to which antedates [Tarski, 2002]. Shortly after defining truth in [Tarski, 1983a], Tarski proves Theorem 3 of the article, on which the consequences of true sentences are true, and the discussion of the ω-rule in § 5 emphasizes the connection between consequence and truth just as [Tarski, 2002] does.

19. I quote the Woodger version since as far as I can tell the relevant sentence in [Tarski, 2002] is ungrammatical.

20. Just how many depends on details about the version of STT assumed, especially whether it includes the axiom of infinity.
21. [Sher, 2008] and her related writings are the best developed presentation of such an approach.
22. The treatment of non-logical expressions as "empirical" predates Carnap in material to which Tarski paid attention, e.g. [Padoa, 1967, 151].
23. One might of course charge Carnap's treatment of Languages I and II with being *ad hoc* itself; see § 7.6.2 below.
24. Compare here [Woleński, 1989, 269] on "the Polish theory of analytic sentences".
25. True to idiosyncratic form, Leśniewski disagreed and maintained that Ontology had no ontological commitments. As [Simons, 1993, 215] notes, the issue turns on the disputed status of the quantifiers in Leśniewski's system that we noted at § 3.1.1.
26. [Sagüillo, 2009, 25] also emphasizes the connection between Tarski's definition and specifically mathematical practice.

8 Conclusion

1. As noted above, this confusion is the source of both of the claim (e.g. [Haack, 1976a, 323–4]) that the T-sentences are compatible with all theories of truth and the claim that the T-sentences are trivial truths that "give no information" (as, e.g., in the criticism of deflationism in [Vision, 2004]).
2. [Ray, 2003] makes something significant out of the difference between this passage and § 1 of [Tarski, 1983a], but in light of [Tarski, 1983a, 267] I don't think the difference comes to much.
3. Some pause here may be given by Carnap's unexplained use of "translates" up to and including the definition of "designation" at [Carnap, 1942, 53–4], but that translation is a matter of sharing designata is made clear at [Carnap, 1942, 53] where Carnap writes "Let us suppose for the moment that we understand a given object language S, say German or S_3 (§ 8), in such a way that we are able to translate its expressions and sentences into the metalanguage M used, say English (including some variables and symbols). It does not matter whether this understanding is based on the knowledge of semantical rules or is intuitive; it is merely supposed that, if an expression is given (say e.g. 'Pferd', 'drei' in German, '$P(a)$' in S_3), for all practical purposes we know an English expression corresponding to it as its 'literal translation'". Here there are two ways to know what translates what—by knowing explicitly stated "semantical rules" or intuitively, but since what is known in the first case is a semantic matter of designation, what is known in the second case must be designation as well. Carnap's appeal to translation in [Carnap, 1942] isn't an appeal to a different conception of meaning, but is rather an implicitly metametalinguistic appeal to relations of sameness of extensional semantic value that is used in judging semantical rules in cases where they aren't purely stipulative, as in his examples of semantics for German and French.

Bibliography

Addison, J. W. (2004). Tarski's theory of definability: Common themes in descriptive set theory, recursive function theory, classical pure logic and finite universe logic. *Annals of Pure and Applied Logic*, 126:77–92.

Ajdukiewicz, K. (1935). Die syntaktische konnexität. *Studia Philosophica*, 1(1–27).

Ajdukiewicz, K. (1966). The logical concept of proof: A methodological essay. *Studia Logica*, 19:12–45.

Ajdukiewicz, K. (1978a). Language and meaning. In *The Scientific World-Perspective and Other Essays 1931–1963*, volume 108 of *Synthese Library*, pages 35–66. Reidel.

Ajdukiewicz, K. (1978b). On the meaning of expressions. In Giedymin, J., editor, *The Scientific World-Perspective and Other Essays 1931–1963*, volume 108 of *Synthese Library*, pages 1–34. Reidel.

Ajdukiewicz, K. (1978c). *The Scientific World-Perspective and Other Essays 1931–1963*, volume 108 of *Synthese Library*. Reidel.

Ajdukiewicz, K. (1978d). Syntactic connection. In Giedymin, J., editor, *The Scientific World-Perspective and Other Essays 1931–1963*, volume 108 of *Synthese Library*, chapter 118–139. Reidel.

Ajdukiewicz, K. (1978e). The world-picture and the conceptual apparatus. In Giedymin, J., editor, *The Scientific World-Perspective and Other Essays 1931–1963*, volume 108 of *Synthese Library*, pages 67–89. Reidel.

Arpaia, S. R. (2006). On Magari's concept of general calculus: Notes on the history of Tarski's methodology of deductive sciences. *History and Philosophy of Logic*, 27(1):9–41.

Awodey, S. and Reck, E. H. (2002). Completeness and categoricity part I: Nineteenth-century axiomatics to twentieth-century metalogic. *History and Philosophy of Logic*, 23:1–30.

Bays, T. (2001). On Tarski on models. *Journal of Symbolic Logic*, 66(4): 1701–1726.

Bellotti, L. (2003). Tarski on logical notions. *Synthese*, 135(3):401–413.

Belnap, N. (1993). On rigorous definitions. *Philosophical Studies*, 72:115–146.

Betti, A. (2004). Leśniewski's early liar, Tarski and natural language. *Annals of Pure and Applied Logic*, 127:267–287.

Betti, A. (2008). Polish axiomatics and its truth: On Tarski's Leśniewskian background and the Ajdukiewicz connection. In Patterson, D., editor, *New Essays on Tarski and Philosophy*, pages 44–71. Oxford University Press.

Black, M. (1948). The semantic definition of truth. *Analysis*, 8(4):49–63.

Blok, W. J. and Pigozzi, D. (1988). Alfred Tarski's work on general metamathematics. *Journal of Symbolic Logic*, 53(1):36–50.

Burge, T. (1984). Semantical paradox. In *Recent Essays on Truth and the Liar Paradox*, pages 83–117. Oxford University Press.

Carnap, R. (1929). *Abriss der Logistik*. Verlag von Julius Springer, Vienna.

Carnap, R. (1942). *Introduction to Semantics*. Harvard University Press.

Carnap, R. (1963). Intellectual autobiography. In Schlipp, P., editor, *The Philosophy of Rudolf Carnap*, pages 3–84. Open Court.

Carnap, R. (2002). *The Logical Syntax of Language*. Open Court.

Casanovas, E. (2007). Logical operations and invariance. *Journal of Philosophical Logic*, 36:33–60.

Coffa, A. (1987). Carnap, Tarski and the search for truth. *Nous*, 21:547–572.

Creath, R. (1999). Carnap's move to semantics: Gains and losses. In Woleński, J. and Köhler, E., editors, *Alfred Tarski and the Vienna Circle: Austro-Polish Connections in Logical Empiricism*, volume 6 of *Vienna Circle Institute Yearbook*, pages 65–76. Kluwer Academic Publishers.

Cummins, R. (1975). The philosophical problem of truth-of. *Canadian Journal of Philosophy*, 5(1):103–122.

Czeżowski, T. (1918). Imiona i zdania. *Przeglkad Filozoficzny*, 21(3-4):101–109.

David, M. (1994). *Correspondence and Disquotation*. Oxford University Press.

David, M. (2008). Tarski's Convention T and the concept of truth. In Patterson, D., editor, *New Essays on Tarski and Philosophy*. Oxford University Press.

Davidson, D. (1990). The structure and content of truth. *Journal of Philosophy*, 87(6):279–328.

Davidson, D. (2002a). Reality without reference. In *Inquiries into Truth and Interpretation*, pages 215–227. Oxford University Press, 2nd edition.

Davidson, D. (2002b). True to the facts. In *Inquiries into Truth and Interpretation*, pages 37–54. Oxford University Press, 2nd edition.

de Rouilhan, P. (2009). Carnap on logical consequence for Languages I and II. In Wagner, P., editor, *Carnap's Logical Syntax of Language*. Palgrave Macmillan, New York.

Detlefsen, M. (2008). Formalism. In Shapiro, S., editor, *The Oxford Handbook of Philosophy of Mathematics and Logic*. Oxford University Press.

Devidi, D. and Solomon, G. (1999). Tarski on "essentially richer" metalanguages. *Journal of Philosophical Logic*, 28:1–28.

Devitt, M. (2001). The metaphysics of deflationary truth. In Schantz, R., editor, *What is Truth?*, pages 60–78. De Gruyter.

Doner, J. and Hodges, W. (1988). Alfred Tarski and decidable theories. *Journal of Symbolic Logic*, 53(1):20–35.

Douglas Patterson. (2005). Learnability and Compositionality. *Mind and Language* 20(3):326–352.

Duda, R. (2004). On the Warsaw interactions of logic and mathematics in the years 1919–1939. *Annals of Pure and Applied Logic*, 127:289–301.

Edwards, J. (2003). Reduction and Tarski's definition of logical consequence. *Notre Dame Journal of Formal Logic*, 44(1):49–62.

Etchemendy, J. (1988). Tarski on truth and logical consequence. *Journal of Symbolic Logic*, 53(1):51–79.

Etchemendy, J. (1990). *The Concept of Logical Consequence*. Harvard University Press.

Etchemendy, J. (2008). Reflections on consequence. In *New Essays on Tarski and Philosophy*. Oxford University Press.

Ewald, W., editor (1996). *From Kant to Hilbert: A Sourcebook in the Foundations of Mathematics*, volume 2. Oxford University Press.

Feferman, A. B. and Feferman, S. (2004). *Alfred Tarski: Life and Logic*. Cambridge University Press.

Feferman, S. (1999). Logic, logics and logicism. *Notre Dame Journal of Formal Logic*, 40(1):31–54.

Feferman, S. (2008). Tarski's conceptual analysis of semantical notions. In Patterson, D., editor, *New Essays on Tarski and Philosophyw*, pages 72–93. Oxford University Press.

Feferman, S. (2010). Set-theoretical invariance criteria for logicality. *Notre Dame Journal of Formal Logic*, 51(1):3–20.

Fernández-Moreno, L. (1997). Truth in pure semantics: A reply to Putnam. *Sorites*, 8:15–23.

Fernández-Moreno, L. (2001). Tarskian truth and the correspondence theory. *Synthese*, 126:123–147.

Ferreirós, J. (1999). *Labyrinth of Thought: A History of Set Theory and it Role in Modern Mathematics*. Birkhaüser Verlag.

Field, H. (1972). Tarski's theory of truth. *Journal of Philosophy*, 69(13):347–375.

Fine, K. and McCarthy, T. (1984). Truth without satisfaction. *Journal of Philosophical Logic*, 13(4):397–421.

Fraenkel, A. (1928). *Einleitung in die Mengenlehre*. Julius Springer, 3rd edition.

Frost-Arnold, G. (2004). Was Tarski's theory of truth motivated by physicalism? *History and Philosophy of Logic*, 25(4):265–280.

García-Carpintero, M. (1996). What is a Tarskian definition of truth? *Philosophical Studies*, 82(2):113–144.

Gödel, K. (1967a). The completeness of the axioms of the functional calculus of logic. In van Heijenoort, J., editor, *From Frege to Gödel: A Source Book in Mathematical Logic, 1879–1931*, pages 582–591. Harvard University Press.

Gödel, K. (1967b). On formally undecidable propositions of *Principia Mathematica* and related systems I. In van Heijenoort, J., editor, *From Frege to Gödel: A Source Book in Mathematical Logic, 1879–1931*, pages 597–616. Harvard University Press.

Gödel, K. (1986). On undecidable propositions of formal mathematical systems. In et al., F., editor, *Collected Works*, volume 1. Oxford University Press.

Goldfarb, W. (1979). Logic in the twenties: The nature of the quantifier. *Journal of Symbolic Logic*, 44(3):351–368.

Gómez-Torrente, M. (1996). Tarski on logical consequence. *Notre Dame Journal of Formal Logic*, 37(1):125–151.

Gómez-Torrente, M. (2002). The problem of logical constants. *The Bulletin of Symbolic Logic*, 8(1):1–37.

Gómez-Torrente, M. (2004). The indefinability of truth in the "wahrheitsbegriff". *Studia Logica*, 126:27–37.

Gómez-Torrente, M. (2009). Rereading Tarski on logical consequence. *Review of Symbolic Logic*, 2(2):249–297.

Haack, S. (1976a). Is it true what they say about Tarski? *Philosophy*, 51(197): 323–336.

Haack, S. (1976b). The pragmatist theory of truth. *British Journal for the Philosophy of Science*, 27(3):231–249.

Hahn, H. (1980). Empiricism, mathematics, and logic. In McGuiness, B., editor, *Empiricism, Logic and Mathematics*, pages 39–42. Reidel.

Hanson, W. H. (1997). The concept of logical consequence. *Philosophical Review*, 106(3):365–409.

Hanson, W. H. (1999). Ray on Tarski on logical consequence. *Journal of Philosophical Logic*, 28 : 607–618.

Hawthorne, J. and Oppy, G. (1997). Minimalism and truth. *Noûs*, 31(2):170–196.

Heck, R. (1997). Tarski, truth and semantics. *Philosophical Review*, 106(4):533–554.

Hempel, C. (1935). On the logical positivists' theory of truth. *Analysis*, 2(4):49–59.

Herzberger, H. (1970). Paradoxes of grounding in semantics. *Journal of Philosophy*, 67(6):145–167.

Hilbert, D. (1971). *Foundations of Geometry*. Open Court.

Hilbert, D. and Ackermann, W. (1928). *Gründzuge der Theoretischen Logik*. Julius Springer, 1st edition.

Hintikka, J. (1996). *Lingua Universalis vs. Calculus Ratiocinator: An Ultimate Presupposition of Twentieth Century Philosophy*. Kluwer Academic Publishers.

Hintikka, J. and Sandu, G. (1999). Tarski's guilty secret: Compositionality. In Woleński, J. and Köhler, E., editors, *Alfred Tarski and the Vienna Circle: Austro-Polish Connections in Logical Empiricism*, volume 6 of *Vienna Circle Institute Yearbook*, pages 217–230. Kluwer Academic Publishers.

Hodges, W. (1986). Truth in a structure. *Proceedings of the Aristotelian Society*, N. S. 86:135–151.

Hodges, W. (2008). Tarski's theory of definition. In Patterson, D., editor, *New Essays on Tarski and Philosophy*, pages 94–132. Oxford University Press.

Hodges, W. (2009). Set theory, model theory and computability theory. In Haaparanta, L., editor, *The Development of Modern Logic*, pages 471–498. Oxford University Press.

Hoffmann, G. A. (2007). The Semantic Theory of Truth: Field's Incompleteness Objection, *Philosophia*, 35:161–170.

Horwich, P. (1982). Three forms of realism. *Synthese*, 51:181–201.

Horwich, P. (2003). A minimalist critique of Tarski on truth. In Jakko Hintikka, Tadeusz Czarnecki, K. K.-P. T. P. and Rojszczak, A., editors, *Philosophy and Logic: In Search of the Polish Tradition*, number 223 in Synthese Library. Kluwer Academic Publishers.

Howard, D. (1996). Relativity, *Eindeutigkeit* and monomorphism: Rudolf Carnap and the development of the categoricity concept in formal semantics. In Giere, R. and Richardson, A., editors, *Origins of Logical Empiricism*, volume XVI of *Minnesota Studies in the Philosophy of Science*, pages 115–164. University of Minnesota Press.

Huntington, E. V. (1902). A complete set of postulates for the theory of absolute continuous magnitude. *Transactions of the American Mathematical Society*, 3(2):264–279.

Jacquette, D. (1994). Tarski's quantificational semantics and meinongian object theory domains. *Pacific Philosophical Quarterly*, 75:88–107.

Jané, I. (2006). What is Tarski's common concept of consequence? *The Bulletin of Symbolic Logic*, 12(1):1–42.

Jennings, R. (1983). Popper, Tarski and relativism. *Analysis*, 43(3):118–123.

Keuth, H. (1978). Tarski's definition of truth and the correspondence theory. *Philosophy of Science*, 45(3):420–430.

Kirkham, R. (1992). *Theories of Truth: A Critical Introduction*. MIT Press.

Kirkham, R. (1993). Tarski's Physicalism. *Erkenntnis*, 38:289–302.

Kneale, W. (1961). Universality and necessity. *British Journal for the Philosophy of Science*, 12(46):89–102.

Kokoszyńska, M. (1936). Über den absoluten wahrheitsbegriff und einige andere semantische begriffe. *Erkenntnis*, 4:143–165.

Kotarbiński, T. (1961). *Elementy Teorii Poznania, Logiki Formalnej i Metodologii Nauk.* Zaklad Narodowy Imienia Ossolińskich–Wydawnictwo.

Kotarbiński, T. (1966). *Gosiology: The Scientific Approach to the Theory of Knowledge.* Pergamon Press.

Kreisel, G. (1967). Informal rigour and completeness proofs. In Lakatos, I., editor, *Problems in the Philosophy of Mathematics*, pages 138–171. North-Holland.

Kripke, S. (1975). Outline of a theory of truth. *Journal of Philosophy*, 72:690–716.

Kripke, S. (1976). Is there a problem about substitutional quantification? In Evans, G. and McDowell, J., editors, *Truth and Meaning*, pages 325–419. Oxford University Press.

Küng, G. (1977). The meaning of the quantifiers in the logic of Leśniewski. *Studia Logica*, pages 309–322.

Langford, C. H. (1926). Some theorems on deducibility. *The Annals of Mathematics*, 28(1/4):16–40.

Lapointe, S. (2011). *Bernard Bolzano's Theoretical Philosophy: An Introduction.* Palgrave Macmillan.

Leeds, S. (1978). Theories of reference and truth. *Erkenntnis*, 13(1):111–129.

Leśniewski, S. (1911). Przyczynek do analizy zdań egzystencjalnych. *Przeglad Filozoficzny*, 14(3):329–345.

Leśniewski, S. (1913). Krytyka logicznej zasady wyllkaczonego środku. *Przeglad Filozoficzny*, 16(3):315–322.

Leśniewski, S. (1929). Grundzüge eines neuen systems der grundlagen der mathematik. *Fundamenta Mathematicae*, 14:1–81.

Leśniewski, S. (1988). *Stanislaw Leśniewski.* Kluwer Academic Publishers.

Leśniewski, S. (1992a). An attempt at a proof of the ontological principle of contradiction. In Surma, Stanislaw, S. J. T. J. B. D. I., editor, *Collected Works*, pages 20–46. Kluwer Academic Publishers.

Leśniewski, S. (1992b). A contribution to the analysis of existential propositions. In Surma, Stanislaw, S. J. T. J. B. D. I., editor, *Collected Works*, pages 1–19. Kluwer Academic Publishers.

Leśniewski, S. (1992c). The critique of the logical principle of the excluded middle. In *Collected Works*, pages 47–85. Kluwer Academic Publishers.

Leśniewski, S. (1992d). Foundations of the general theory of sets. In *Collected Works*, pages 129–173. Kluwer Academic Publishers.

Leśniewski, S. (1992e). Fundamentals of a new system of the foundations of mathematics. In *Collected Works*, pages 410–605. Kluwer Academic Publishers.

Leśniewski, S. (1992f). Introductory remarks to the continuation of my article: 'grundzüge eines neuen systems der grundlagen der mathematik'. In *Collected Works*, pages 649–710. Kluwer Academic Publishers.

Leśniewski, S. (1992g). Is the class of classes not subordinated to themselves, subordinated to itself? In *Collected Works*, pages 115–128. Kluwer Academic Publishers.

Leśniewski, S. (1992h). On the foundations of mathematics. In *Collected Works*, pages 174–382. Kluwer Academic Publishers.

Levison, A. B. (1965). Logic, language, and consistency in Tarski's theory of truth. *Philosophy and Phenomenological Research*, 25(3):384–392.

Levy, A. (1988). Alfred Tarski's work in set theory. *Journal of Symbolic Logic*, 53(1): 2–6.

Lewis, D. (1969). *Convention: A Philosophical Study*. Blackwell.

Lewis, D. (1970). General semantics. *Synthese*, 22(1–2):18–67.

Löwenheim, L. (1967). On possibilities in the calculus of relatives. In van Heijenoort, J., editor, *From Frege to Gödel: A Source Book in Mathematical Logic, 1879-1931*, pages 228–251. Harvard University Press.

Łukasiewicz, J. (1970a). Logical foundations of probability theory. In Borkowski, L., editor, *Selected Works*, pages 16–63. North-Holland.

Łukasiewicz, J. (1970b). On the concept of magnitude. In Borkowski, L., editor, *Selected Works*, pages 64–83. North-Holland.

Mancosu, P. (2005). Harvard 1940–41: Tarski, Carnap and Quine on a finitistic language of mathematics for science. *History and Philosophy of Logic*, 26: 327–357.

Mancosu, P. (2006). Tarski on models and logical consequence. In Ferreirós, J. and Gray, J. J., editors, *The Architecture of Modern Mathematics: Essays in History and Philosophy*, pages 209–237. Oxford University Press.

Mancosu, P. (2008). Tarski, Neurath and Kokoszyńska on the semantic conception of truth. In Patterson, D., editor, *New Essays on Tarski and Philosophy*, pages 192–224. Oxford University Press.

Mancosu, P., Zach, R., and Badesa, C. (2009). The development of mathematical logic from Russell to Tarski, 1900–1935. In Haaparanta, L., editor, *The Development of Modern Logic*, pages 318–470. Oxford University Press.

Martin, R. L. (1976). Are natural languages universal? *Synthese*, 32(3/4).

McDowell, J. (1978). Physicalism and primitive denotation: Field on Tarski. *Erkenntnis*, 13(1):131–152.

McGee, V. (1992). Two problems with tarski's theory of consequence. *Proceedings of the Aristotelian Society, New Series*, 92:273–292.

McGee, V. (1996). Logical operations. *Journal of Philosophical Logic*, 25:567–580.

McNulty, G. F. (1986). Alfred Tarski and undecidable theories. *Journal of Symbolic Logic*, 51(4):890–898.

Milne, P. (1997). Tarski on truth and its definition. In Childers, K. and Svoboda, editors, *Logica '96: Proceedings of the 10th Internatonal Symposium*, pages 189–210. Filosofia, Prague.

Milne, P. (1999). Tarski, truth and model theory. *Proceedings of the Aristotelian Society*, 99:141–167.

Monk, J. D. (1986). The contributions of Alfred Tarski to algebraic logic. *Journal of Symbolic Logic*, 51(4):899–906.

Mormann, T. (1999). Neurath's opposition to Tarskian semantics. In Woleński, J. and Köhler, E., editors, *Alfred Tarski and the Vienna Circle: Austro-Polish Connections in Logical Empiricism*, volume 6 of *Vienna Circle Institute Yearbook*, pages 165–178. Kluwer Academic Publishers.

Mou, B. (2001). The enumerative character of Tarski's definition of truth and its general character in a Tarskian system. *Synthese*, 126:91–121.

Murawski, R. (2004). Philosophical reflection on mathematics in poland in the interwar period. *Annals of Pure and Applied Logic*, 127:325–337.

Niiniluoto, I. (1999a). Tarskian truth as correspondence — replies to some objections. In Peregrin, J., editor, *Truth and Its Nature (If Any)*, pages 91–104. Kluwer Academic Publishers.

Niiniluoto, I. (1999b). Theories of truth: Vienna, Berlin and Warsaw. In Woleński, J. and Köhler, E., editors, *Alfred Tarski and the Vienna Circle: Austro-Polish Connections in Logical Empiricism*, volume 6 of *Vienna Circle Institute Yearbook*, pages 17–26. Kluwer Academic Publishers.

Padoa, A. (1967). Logical introduction to any deductive theory. In van Heijenoort, J., editor, *From Frege to Gödel: A Source Book in Mathematical Logic, 1879–1931*, pages 118–123. Harvard University Press.

Pasch, M. (1912). *Vorlesungen über neuere Geometrie*. B. G. Teubner, Leipzig.

Patterson, D. (2002). Theories of truth and Convention T. *Philosophers' Imprint*, 2(5):1–16.

Patterson, D. (2003). What is a correspondence theory of truth? *Synthese*, 137(3):421–444.

Patterson, D. (2005). Deflationism and the truth conditional theory of meaning. *Philosophical Studies*, 124(3):271–294.

Patterson, D. (2006). Tarski, the liar and inconsistent languages. *The Monist*, 89(1):150–177.

Patterson, D. (2007). On the determination argument against deflationism. *Pacific Philosophical Quarterly*, 88(2):243–250.

Patterson, D. (2008a). Tarski's conception of meaning. In Patterson, D., editor, *New Essays on Tarski and Philosophy*, pages 157–191. Oxford University Press.

Patterson, D. (2008b). Truth definitions and definitional truth. *Midwest Studies in Philosophy*, 32:313–328.

Patterson, D. (2008c). Understanding the liar. In Beall, J. C., editor, *Revenge of the Liar: New Essays on the Paradox*, pages 197–224. Oxford University Press.

Patterson, D. (2009). Inconsistency theories of semantic paradox. *Philosophy and Phenomenological Research*, 79(2):387–422.

Perkins, M. (1952). Notes on the pragmatic theory of truth. *Journal of Philosophy*, 49(18):573–587.

Pieri, M. (1901). Sur la géométrie envisagée comme un système purement logique. In *Bibliothèque du Congrès International de Philosophie, Paris 1900*, pages 235–272. A. Colin, Paris.

Popper, K. (1963). *Conjectures and Refutations: The Growth of Scientific Knowledge*. Routledge.

Popper, K. (1974). Some philosophical comments on Tarski's theory of truth. In et al., L. H., editor, *Proceedings of the Tarski Symposium*. American Mathematical Society.

Popper, K. (2002). *The Logic of Scientific Discovery*. Routledge.

Priest, G. (1995). Etchemendy on logical consequence. *Canadian Journal of Philosophy*, 25:283–292.

Putnam, H. (1975). Do true assertions correspond to reality? In *Mind, Language and Reality: Philosophical Papers*, volume 2. Cambridge University Press.

Putnam, H. (1994). A comparison of something with something else. In Conant, J., editor, *Words and Life*, pages 330–350. Harvard University Press.

Raatikainen, P. (2003). More on Putnam on Tarski. *Synthese*, 135:37–47.

Raatikainen, P. (2008). Truth, meaning, and translation. In Patterson, D., editor, *New Essays on Tarski and Philosophy*, pages 247–262. Oxford University Press.

Ray, G. (1996). Logical consequence: A defense of Tarski. *Journal of Philosophical Logic*, 25:617–677.

Ray, G. (2003). Tarski, Soames and the metalinguistic liar. *Philosophical Studies*, 115:55–80.

Ray, G. (2005). On the matter of essential richness. *Journal of Philosophical Logic*, 34:433–457.

Reck, E. H. (2004). From Frege and Russell to Carnap: Logic and logicism in the 1920s. In Awodey, S. and Klein, C., editors, *Carnap Brought Home: The View From Jena*, pages 151–180. Open Court Publishing.

Richard, J. (1967). The principles of mathematics and the problem of sets. In van Heijenoort, J., editor, *From Frege to Gödel: A Source Book in Mathematical Logic, 1879–1931*, pages 142–144. Harvard University Press.

Ricketts, T. (2004). Frege, Carnap and Quine: Continuities and discontinuities. In Awodey, S. and Klein, C., editors, *Carnap Brought Home: The View From Jena*, pages 181–202. Open Court.

Ricketts, T. (2009). From tolerance to reciprocal containment. In Wagner, P., editor, *Carnap's Logical Syntax of Language*, pages 217–235. Palgrave Macmillan.

Rogers, R. (1963). A survey of formal semantics. *Synthese*, 15(1):17–56.

Rojszczak, A. (1999). Why should a physical object take on the role of a truth-bearer? In Wolenski, J. and Kohler, E., editors, *Alfred Tarski and the Vienna Circle*. Kluwer Academic Publishers.

Rojszczak, A. (2002). Philosophical background and philosophical content of the semantic definition of truth. *Erkenntnis*, 56:29–62.

Russell, B. (1903). *The Principles of Mathematics*. W. W. Norton.

Sagüillo, J. (1997). Logical consequence revisited. *The Bulletin of Symbolic Logic*, 3(2):216–241.

Sagüillo, J. (2009). Methodological practice and complementary concepts of logical consequence: Tarski's model-theoretic consequence and Corcoran's information-theoretic consequence. *History and Philosophy of Logic*, 30(1):21–48.

Scanlan, M. (1991). Who were the American postulate theorists? *The Journal of Symbolic Logic*, 56(3):981–1002.

Scanlan, M. (2003). American postulate theorists and Alfred Tarski. *History and Philosophy of Logic*, 24:307–325.

Schantz, R. (1998). Was Tarski a deflationist? *Logic and Logical Philosophy*, pages 157–172.

Sher, G. (1991). *The Bounds of Logic: A Generalized Viewpoint*. MIT Press.

Sher, G. (1996). Did Tarski commit Tarski's Fallacy? *The Journal of Symbolic Logic*, 61(2):653–686.

Sher, G. (1999). On the possibility of a substantive theory of truth. *Synthese*, 117:133–172.

Sher, G. (2004). In search of a substantive theory of truth. *Journal of Philosophy*, 101(1):5–36.

Sher, G. (2008). Tarski's thesis. In Patterson, D., editor, *New Essays on Tarski and Philosophy*, pages 300–339. Oxford University Press.

Simmons, K. (1993). *Universality and the Liar: An Essay on Truth and the Diagonal Argument*. Cambridge University Press.

Simons, P. (1993). Nominalism in Poland. In Francesco Coniglione, R. P. and Wolenski, J., editors, *Polish Scientific Philosophy: The Lvov–Warsaw School*, volume 28 of *Poznań Studies in the Philosophy of the Sciences and the Humanities*, pages 207–31. Rodopi.

Simons, P. (2009). Twardowski on truth. *The Baltic International Yearbook of Cognition, Logic and Communication*, 4:1–14.

Sinaceur, H. (2001). Alfred Tarski: Semantic shift, heuristic shift in metamathematics. *Synthese*, 126:49–65.

Sinisi, V. F. (1967). Tarski on the inconsistency of colloquial language. *Philosophy and Phenomenological Research*, 27(4):537–541.

Skolem, T. (1967). Some remarks on axiomatized set theory. In van Heijenoort, J., editor, *From Frege to Gödel: A Source Book in Mathematical Logic, 1879-1931*, pages 290–301. Harvard University Press.

Sloman, A. (1971). Tarski, Frege and the liar paradox. *Philosophy*, 46(176): 133–147.

Sluga, H. (1999). Truth before Tarski. In Woleński, J. and Köhler, E., editors, *Alfred Tarski and the Vienna Circle: Austro-Polish Connections in Logical Empiricism*, volume 6 of *Vienna Circle Institute Yearbook*. Kluwer Academic Publishers.

Smith, B. (1994). *Austrian Philosophy: The Legacy of Franz Brentano*. Open Court.

Soames, S. (1984). What is a theory of truth? *Journal of Philosophy*, 81(8):411–429.

Soames, S. (1999). *Understanding Truth*. Oxford University Press.

Stroińska, M. and Hitchcock, D. (2002). On the concept of following logically, translator's introduction. *History and Philosophy of Logic*, 23:155–175.

Sundholm, G. (2003). Tarski and Leśniewski on languages with meaning versus languages without use. In J. Hintikka, K. Czarnecki, K. K.-P. T. P. A. R., editor, *Philosophy and Logic: In Search of the Polish Tradition*, pages 109–128. Kluwer Academic Publishers.

Suppes, P. (1988). Philosophical implications of Tarski's work. *Journal of Symbolic Logic*, 53(1):80–91.

Tarski, A. (1933). *Pojḱecie Prawdy w Jḱezykach Nauk Dedukcyjnych*. NakŁadem Towarstwa Naukowego Warsawskiego.

Tarski, A. (1983a). The concept of truth in formalized languages. In Woodger, J. H., editor, *Logic, Semantics, Metamathematics: Papers from 1923 to 1938*, pages 152–278. Hackett Publishing.

Tarski, A. (1983b). The establishment of scientific semantics. In Woodger, J. H., editor, *Logic, Semantics, Metamathematics: Papers from 1923 to 1938*, pages 401–408. Hackett Publishing.

Tarski, A. (1983c). Fundamental concepts of the methodology of the deductive sciences. In Woodger, J. H., editor, *Logic, Semantics, Metamathematics: Papers from 1923 to 1938*, pages 60–109. Hackett Publishing, 2nd edition.

Tarski, A. (1983d). Investigations into the sentential calculus. In Woodger, J. H., editor, *Logic, Semantics, Metamathematics: Papers from 1923 to 1938*, pages 38–59. Hackett Publishing, 2nd edition.

Tarski, A. (1983e). *Logic, Semantics, Metamathematics: Papers from 1923 to 1938*. Hackett Publishing, Indianapolis, 2nd edition.

Tarski, A. (1983f). On definable sets of real numbers. In Woodger, J. H., editor, *Logic, Semantics, Metamathematics: Papers from 1923 to 1938*, pages 110–142. Hackett Publishing.

Tarski, A. (1983g). On some fundamental concepts of metamathematics. In Woodger, J. H., editor, *Logic, Semantics, Metamathematics: Papers from 1923 to 1938*, pages 30–37. Hackett Publishing, 2nd edition.

Tarski, A. (1983h). On the concept of logical consequence. In Tarski (1983e).

Tarski, A. (1983i). On the limitations of the means of expression of deductive theories. In Woodger, J. H., editor, *Logic, Semantics, Metamathematics: Papers from 1923 to 1938*, pages 384–392. Hackett Publishing.

Tarski, A. (1983j). On the primitive term of logistic. In Woodger, J. H., editor, *Logic, Semantics, Metamathematics: Papers from 1923 to 1938*, pages 1–23. Hackett Publishing.

Tarski, A. (1983k). Some methodological investigations on the definability of concepts. In *Logic, Semantics, Metamathematics: Papers from 1923 to 1938*, pages 298–319. Hackett Publishing, 2nd edition.

Tarski, A. (1983l). Some observations of the concepts of ω-consistency and ω-completeness. In Woodger, J. H., editor, *Logic, Semantics, Metamathematics: Papers from 1923 to 1938*, pages 279–295. Hackett Publishing.

Tarski, A. (1986a). Der warheitsbegriff in den formalisierten sprachen. In Givant, S. R. and MacKenzie, R. N., editors, *Collected Papers*, volume 2, pages 58–198. Birkhauser.

Tarski, A. (1986b). Einige methodologische untersuchungen über die definierbarkeit der begriffe. In Rota, G.-C., editor, *Collected Papers*, volume 1, pages 639–659. Birkhauser.

Tarski, A. (1986c). Gründlegung der wissenschaftlichen semantik. In Rota, G.-C., editor, *Collected Works*, volume 2, pages xxx–xxx. Birkhauser.

Tarski, A. (1986d). On undecidable statements in enlarged systems of logic and the concept of truth. In Givant, S. R., editor, *Collected Works*, pages 559–568. Birkhaüser Verlag.

Tarski, A. (1986e). The semantic conception of truth and the foundations of semantics. In Rota, G.-C., editor, *Collected Papers*, volume 2, pages 661–699. Birkhauser.

Tarski, A. (1986f). Sur la méthode déductive. In Rota, G.-C., editor, *Collected Papers*, volume 2, pages 323–333. Birkhauser.

Tarski, A. (1986g). Sur les ensembles définissables de nombres réels i. In Rota, G.-C., editor, *Collected Papers*, volume 2, pages 519–548. Birkhauser.

Tarski, A. (1986h). Über die beschränktheit der ausdrucksmittle deduktiver theorien. In Rota, G.-C., editor, *Collected Papers*, volume 2, pages 203–212. Birkhauser.

Tarski, A. (1986i). Über einige fundamentalen begriffe der metamathematik. In Rota, G.-C., editor, *Collected Papers*, volume 1, pages 311–320. Birkhauser.

Tarski, A. (1986j). What are logical notions? *History and Philosophy of Logic*, 7: 143–154.

Tarski, A. (1992). Alfred Tarski: Drei briefe an Otto Neurath. *Grazer Philsophische Studien*, 43:1–32.

Tarski, A. (1995). *Introduction to Logic and to the Methodology of Deductive Sciences*. Dover.

Tarski, A. (2002). On the concept of following logically. *History and Philosophy of Logic*, 23(3):155–196.

Tharp, L. (1971). Truth, quantification and abstract objects. *Nous*, 5:363–372.

Twardowski, K. (1977). *On the Content and Object of Presentations*, volume 4 of *Melbourne International Philosophy Series*. Martinus Nijhoff.

van den Dries, L. (1988). Alfred Tarski's elimination theory for real closed fields. *Journal of Symbolic Logic*, 53(1):7–19.

van Heijenoort, J. (1967). Logic as calculus vs logic as language. *Synthese*, 17: 324–330.

van Orman Quine, W. (1960). *Word and Object*. MIT Press.

van Orman Quine, W. (1966). Truth by convention. In *The Ways of Paradox and Other Essays*. Random House.

van Orman Quine, W. (1973). *The Roots of Reference*. Open Court Publishing.

Vaught, R. L. (1974). Model theory before 1945. In L. Henkin, J. Addison, C. C. C. W. C. D. S. R. V., editor, *Proceedings of the Tarski Symposium*. American Mathematical Society.

Vaught, R. L. (1986). Alfred Tarski's work in model theory. *The Journal of Symbolic Logic*, 51(4):869–882.

Veblen, O. (1904). A system of axioms for geometry. *Transactions of the American Mathematical Society*, 5(3):343–384.

Veblen, O. (1906). The foundations of geometry: A historical sketch and a simple example. *Popular Science Monthly*, 68:21–28.

Vision, G. (2004). *Veritas: The Correspondence Theory and Its Critics*. MIT Press.

Wallace, J. (1970). On the frame of reference. *Synthese*, 22(1/2):117–150.

Weingartner, P. (1999). Tarski's truth condition revisited. In Woleński, J. and Köhler, E., editors, *Alfred Tarski and the Vienna Circle: Austro-Polish Connections in Logical Empiricism*, volume 6 of *Vienna Circle Institute Yearbook*, pages 193–201. Kluwer Academic Publishers.

White, M. and Tarski, A. (1987). A philosophical letter of Alfred Tarski. *Journal of Philosophy*, 84(1):28–32.

Woleński, J. (1989). *Logic and Philosophy in the Lvov–Warsaw School*, volume 198 of *Synthese Library*. Kluwer Academic Publishers.

Woleński, J. (1994). Jan Łukasiewicz on the liar paradox, logical consequence, truth and induction. *Modern Logic*, 4(4):392–400.

Woleński, J. (1999). Semantic revolution: Rudolf Carnap, Kurt Gödel, Alfred Tarski. In Eckehart Köhler, J. W., editor, *Alfred Tarski and the Vienna Circle*, volume 6 of *Vienna Circle Institute Yearbook*, pages 1–15. Kluwer Academic Publishers.

Woleński, J. and Simons, P. (1989). De veritate: Austro-polish contributions to the theory of truth from Brentano to Tarski. In Szaniawski, K., editor, *The Vienna Circle and the Lvov–Warsaw School*, pages 391–442. Kluwer Academic Publishers.

Wolenski, J. and Murawski, R. (2008). Tarski and his Polish Predecessors on Truth. In Douglas Patterson, editor, *New Essays on Tarski and Philosophy*, pages 21–43. Oxford University Press.

Wybraniec-Skardowska, U. (2004). Foundations for the formalization of meta-mathematics and axiomatizations of consequence theories. *Annals of Pure and Applied Logic*, 127:243–266.

Index

Printed in the United States
By Bookmasters